Medusa of Time: How Technology Redefines Copenhagen

by
T. Lee Baumann

1st Printing August 2012
Revised Edition March 2013

CreateSpace
7290-B Investment Drive
North Charleston, SC 29418, U.S.A

Baumann, T. Lee, 1950-
        Medusa of Time: How Technology
        Redefines Copenhagen
        by T. Lee Baumann

Includes endnotes and index.

    ISBN-10: 1479167398
    ISBN-13: 978-1479167395
    LCCN: 2012916017

This book is dedicated to all the physics giants...upon whose shoulders we teeter.

The author is grateful for permission to use excerpts from the following works:

Merriam-Webster's Collegiate Dictionary, Tenth Edition. By permission. *Merriam-Webster's Collegiate® Dictionary*, Tenth Edition ©2000 by Merriam-Webster, Inc.

Table of Contents

Foreword

Physicists recognize that a photon or light particle travels at the finite speed of 186,000 miles per second (300,000 km/sec). Yet, quantum mechanics, thanks to the genius of Albert Einstein in his special theory, states that time stops at the speed of light and imbues the light wave with infinite speed. Although Einstein's proposal existed as theory back in 1905, it exists as proven and accepted fact today. These two differing—but proven—speeds for the mysterious photon illustrate one of the obvious conundrums within the field of quantum mechanics. Hence, physicists question why it is that at one moment they measure the photon at 186,000 mi/sec and at another, infinity.

This book will argue that this discrepancy exists, simply by determining whether the said observation (and speed) was made within the dimension of time. My research indicates that if the photon or light wave is measured within the dimension of time, it is observed to travel at 186,000 mi/sec—as a particle. The world in which we live is a material (particulate) one. When the photon's behavior is observed as occurring outside of time—as a wave—scientists establish that it is everywhere at once (infinite speed). [1]

We know that the dimension of time is real. We experience it every moment of our lives. We, as humans, are trapped within the dimension of time. Yet, science experiments have allowed us to expose light's ability to exist outside of time. Through science, we know that the mysterious photon may exist both within and outside of time—just as Einstein's special theory predicted.

This text will outline how this observation is equally true for particulate and macroscopic objects of any size. Science experiments have already shown that this holds true for microscopic objects like electrons, atoms, and even large molecules. In 1924, Louis de Broglie demonstrated that the same must hold true for macroscopic objects of all sizes. Every particulate or material object has an associated wave—just like the photon. For his research, de Broglie won the Nobel Prize in 1929.

By extending de Broglie's corollary to the dimension of time, I realized that all waves (whether of microscopic or macroscopic objects) must also exist outside of time—just like the light wave. Similarly, all particles (whether of microscopic or macroscopic objects) must also exist within time.

Our very existence, restricted to the dimension of time, dictates that we can experience our reality only as defined by particles and matter. Even when we measure the existence of waves in scientific experiments (such as through wave interference), we do so retroactively and by observing the waves' previous behavior. We actually measure this past behavior, however, as particles.

There is no longer the need to demonstrate that the Higgs boson or "God particle" renders the property of mass to particles. I contend, rather, that it is the dimension of time.

In 1990, Itano, et. al. conducted a classic experiment at the National Institute of Standards and Technology (NIST)[2] which represented a major advance in the worldview of the Copenhagen Interpretation of quantum physics.

The Copenhagen Interpretation views the wave-particle nature of any object as a combination of both wave *and* particle...until an experimenter performs a measurement or observation. At the moment of the measurement, something unusual occurs that defines whether the object was a wave or a particle. Although this statement may not sound very profound, it would take a good many years for many physicists to understand its full implications.

The claim that the mere act of human observation could alter the progress of a scientific experiment certainly caught my attention. The NIST experiment represented just one additional, "impossible" facet involving the mysterious field of quantum physics and the role that observation has upon our four-dimensional reality.

Some physicists believe that human consciousness is the critical element in the observation process that defines whether an object is measured as a wave or a particle. Many just plead a factual, "Who knows?" As already noted, my research indicates that an entirely unrelated element is the critical defining factor for whether something exists as a wave or particle: the ill-defined dimension of *time*.

As I reflect upon the first sentences of this book, I recall how my first attempts at processing even the basic elements of quantum physics had a learning curve of many years, beginning for me in the 1970s. By the year 1999, I had written the first rough draft of my first book, *God at the Speed of Light*, which dealt with the supernatural ramifications of this relatively new field.[3]

As *God at the Speed of Light* demonstrated, the metaphysical ramifications of quantum mechanics nicely complement and reinforce the traditional views of the world's major religions. Science no longer opposes spirituality—but supports it.

Back in the mid-1970s, I was an atheist because of science. Today I stand as a relentless advocate for the existence of a Higher Power for the same reason. Science has always been my truest compass. Yet, I remain astonished that the field now supports the very existence of that which I once thought impossible: God.

My spiritual evolution has taken me through a fulfilling and rewarding vocational cycle from atheist-physician to believer to writer to cruise-ship "enrichment lecturer." Through my travels, I have met the most fascinating people—from all walks of life, cultures, and economies of scale. I continue to learn from these remarkable personal interactions and consider myself the richer for it.

God has placed our human race within this unique four-dimensional realm for a reason. Learning, growing, making mistakes, and adaptation are all part of the process.

We have witnessed this type of growth

cycle in the civil rights struggles in the United States and in the apartheid era in South Africa. Yet, despite these conflicts, our human race is learning and changing for the better. It is a slow and painful process, but there is reason for hope. I can offer one personal example of my optimism from when I was in South Africa (where I spent a month volunteering) in 2012. My Afrikaans (language) instructor told me that it was considered rude if you did *not* say "hello" to *everyone* you passed on the streets of Cape Town. I immediately sought clarification. "Everyone?" I asked incredulously, aware of how uncommon this practice was in the United States.

"Yes, everyone," was the unequivocal response.

So, I did the test. During my next walk into Rondebosch, a popular shopping area outside the city, I made a point of saying "hello" to everyone I passed on the street. I was astonished. Despite the past history of oppression from apartheid and the abject poverty of many of these proud people, *everyone* made eye contact, smiled, and responded with a "hello" or "molo." I should point out that there *were* exceptions—the tourists! Few foreigners responded in kind. As I retell the story, "most tourists looked at me like I had two heads." I was still learning.

There are reasons for hope in our future, and education is playing a well-recognized, crucial role. Our rapidly-advancing technology is catalyzing an ever-expanding growth in education for anyone with access to the Internet. I am especially impressed by the rapid expansion in scientific knowledge of the various people I have met around the world—most notably the children. The average person in this modern, Internet-connected world is smarter than I ever was at the corresponding age. Recently at a "trivia of our solar system" lecture I conducted, a young, 9-year-old Hispanic girl accurately answered that Venus was one of our solar system's two planets that spins in a direction opposite to that of Earth. The other is Uranus. How many people know that?

At that age, I barely knew what a planet was. I have no doubt that this young girl *knew* the answer. The only previous response to my question had been the alternate correct answer of Uranus—by a well-educated adult. This young girl had tugged at her mother's blouse for a translation of my question and, again, to convey her response.

Perhaps—and hopefully—science is making a comeback in our schools. No one doubts that our children hold the key to the future.

As you might guess, I am a staunch proponent of education. From the sales of my books, _all_ proceeds go to any one of my *God at the Speed of Light* college scholarships throughout the southeastern United States. Next to my family, these scholarships represent my proudest achievement.

If this book stimulates, in any small way, added interest in the field of science, I will have achieved my goal.

frequently clashed with school authorities and failed several crucial entrance examinations for secondary school (1895). Yet, he always excelled in his beloved physics and mathematics.

One year later (1896), Einstein finally passed the necessary entrance exams and entered a four-year mathematics and physics program at the Zurich Polytechnic.

It was at the Polytechnic that Einstein met his first wife, Mileva Marić. At the time of their marriage (1903), Albert had started working for the Federal Office for Intellectual Property, the well-known patent office in Bern, Switzerland.

During his time at the patent office, Einstein confronted various auspicious problems dealing with the propagation of electricity and the synchronization of time—problems that he would successfully and famously overcome.

The next two years proved unusually stimulating and creative for Einstein. In 1905, often dubbed his "miracle year," Einstein completed his thesis on "A New Determination of Molecular Dimensions" and published four monumental papers (any one of which would have garnered him lasting success) on the photoelectric effect, Brownian motion, special relativity, and the equivalence of matter and energy.

In 1907, Einstein published his first of several papers on general (think of "**g**" for gravity) relativity. It would not be until 1919 that Sir Arthur Eddington would lead an expedition that would uphold Einstein's groundbreaking theory on gravity. The expedition proved that the Sun's gravity could, just as Einstein predicted, bend the path of light from other stars. The newspaper headlines in Britain declared, "Revolution in Science—New Theory of the Universe—Newtonian Ideas Overthrown."

Within two years (1921), the Nobel Prize committee awarded Einstein its prestigious prize in physics for, surprisingly, his explanation of the photoelectric effect. The Nobel Prize committee still regarded his theories on special and general relativity as too controversial to consider.

In 1940, Einstein became a United States citizen.

At age 76, on April 17, 1955, Albert Einstein died from internal bleeding caused by the rupture of an abdominal aortic aneurysm. Einstein refused surgery on the vessel, stating, "I want to go when I want. It is tasteless to prolong life artificially. I have done my share; it is time to go. I will do it elegantly."

All things considered, most Americans would classify Einstein as the greatest physicist who ever lived.

*   *   *

Werner Karl Heisenberg was born in Würzburg, Germany in 1901. His father, a secondary school teacher, could never have imagined how this young family addition would drastically alter the landscape of the known physics world.

Young Werner would study mathematics and physics at the Universities of Munich and Göttingen from 1920-23. In 1922, he attended a local festival in Göttingen where he met Niels Bohr for the first time. The meeting would make a lasting impression and launch a life-long partnership.

By 1924, Heisenberg was performing research with Bohr at the University of

Copenhagen. In 1926, Heisenberg graduated to university lecturer and was working as Bohr's assistant.

One year later (1927), Heisenberg formulated the uncertainty principle that would make him famous. It states that no measurement can *simultaneously* determine both a wave/particle's position and momentum. During this time, Heisenberg also found himself collaborating with Bohr and Einstein to put together their first thoughts on quantum mechanical theory.

In 1929, Heisenberg gave a series of lectures at the University of Chicago, during which he described this new field of quantum mechanics.

Historically, the "Copenhagen Interpretation" got its name from a notation in the preface to his book, *The Physical Principles of the Quantum Theory* (1930). In the preface, Heisenberg recorded:

On the whole the book contains nothing that is not to be found in previous publications, particularly in the investigations of Bohr. The purpose of the book seems to me to be fulfilled if it contributes somewhat to the diffusion of

that   'Kopenhagener   Geist   der
Quantentheorie' [i.e., Copenhagen spirit
of quantum theory]....

By 1932, quantum mechanics had
transitioned to an accepted construct within
the field of physics for which Heisenberg
would receive the Nobel Prize in physics
the same year. History records that the
accolade was "for the creation of quantum
mechanics, the application of which...led
to the discovery of the allotropic forms of
hydrogen."[4]

                    *   *   *

Danish physicist, **Niels Bohr**, won the
Nobel Prize in physics in 1922—one year
after Einstein received his. The award was
for his "services in the investigation of the
structure of atoms and of the radiation
emanating from them."[5]

Having been born and raised in
Copenhagen, Bohr founded the Institute of
Theoretical Physics there in 1921 and
became its director.

Although Bohr is most widely known
for his Bohr model of the atom that many

of us studied as students, he also played a major role along with Heisenberg, Einstein, and others in developing quantum mechanics. Bohr's Institute of Theoretical Physics would, in fact, serve as a popular gathering place for many of the field's early pioneers during the 1920s and 30s.

The final product

The first formulations of quantum mechanics date back to the early 1900s. In 1900, physicist Max Planck discovered that the electromagnetic radiation of black bodies occurred only in discrete energy packets or "quanta," h·f, where h is his own Planck's constant[6] and f is the frequency. This early finding suggested that energy could be viewed as possessing small, indivisible, single units of measure. This naïve belief would not last, and the long-held dividing line between energy and matter would also quickly become blurred.

In a little more than a decade (1913), Niels Bohr produced his first model of the basic building block of matter: the atom— along with its well-defined electron orbits.

Yet, Bohr's model had problems, and he and others constantly worked to enhance it.

In 1925, Werner Heisenberg formulated the basic elements of a new model which would emerge into what we better recognize today as quantum mechanics.

As quantum physics continued to evolve, more complicated problems arose. One complex issue pertained to whether certain "quanta of energy" existed as waves or particles. Physicists soon identified that a particle could not simultaneously act as a wave. A particle was a particle, and a wave was a wave. A wave could transform into a particle—and vice versa, but one could not act as both at the same time. One was mutually exclusive of the other, but each shared equal importance—and each was real. Hence, the term "wave-particle duality" developed to describe this schizophrenic nature of waves and particles. In one experiment, an atom, electron, or photon would act like a wave. In another, each would act like a particle.

Bohr viewed the wave-particle nature of any object as a combination of the two (i.e., wave *and* particle) until the experimenter performed an actual measurement. The researcher might observe the object of their attention as either a wave or a particle, depending upon the circumstances and how the experiment was designed. This paragraph defines the essence of the early version of the Copenhagen Interpretation.

Most scientists now appreciate that, at the moment of observation, something unusual occurs that transforms an object's *wave function* into a well-defined, particulate object. The wave function is an all-inclusive definition of every conceivable state of an object.

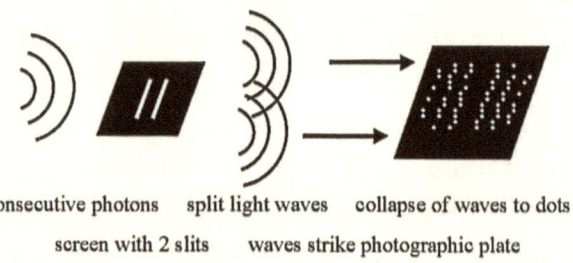

consecutive photons   split light waves   collapse of waves to dots

screen with 2 slits   waves strike photographic plate

Figure 1.1 The double-slit experiment

For instance, in the double-slit experiment **(Figure 1.1)**, *consecutively-fired, single light waves* split into two "half" or "progeny" photons in the middle of the experiment (possible, by the way, only for *waves*, not particles). Each pair of split waves subsequently collapse into a single light particle at the end of the experiment to form a single dot on a photographic plate (the site of "observation" or measurement).

As the dots accumulate or coalesce onto the photographic plate (from the consecutively fired light waves), the resultant image gradually reveals an *interference pattern* for light—the classic indication that the pairs of split photons existed previously as waves in their unobserved states (*prior* to collapsing to particles onto the plate).

Redefining the Copenhagen Interpretation

One version of the Copenhagen Interpretation states that the sole act of measuring, observing, seeing, or otherwise "sensing" (hearing, smelling, tasting, or touching) any collection of waves will cause them to "collapse" into their recognizable, solid forms. A tree, an electron, visible light—you name it—can all exist as waves before becoming material, particulate objects.

Even though it may seem nonsensical to describe a tree as a wave, quantum physics—as we shall see—does exactly that. Before examining how macroscopic objects can exist as waves however, let's look at what mechanisms might explain how a wave can possibly transform to a particle.

As previously noted, many scientists believe that consciousness might be the magical catalyst that somehow causes the transition from wave to particle—called, in physics parlance, the "collapse of the wave function." One oft-quoted explanation is as follows:

[Mathematician John] von Neumann could not find a natural place to locate [the collapse of the wave function] "miracle." Everything, after all, is made of atoms: there's nothing holy about a measuring instrument. Following the von Neumann chain, driven by his own logic, in desperation von Neumann seized on its only peculiar link: the process by which a physical signal in the brain becomes an experience...human consciousness is the site of the wave function collapse.[7]

Von Neumann is not speaking abstractly here. Rather, in the absence of any human presence, solid matter (a table, for instance) does *not* exist as we know it. Instead, the object exists as an ill-defined transmutation of Einstein's "E," from his equation $E = mc^2$. Imagine walking through a brick wall the same as walking through a room permeated by our present-day electromagnetic (e.g., cell phone, TV, and radio) waves.

Many of us have heard the proverbial question regarding the tree falling in a forest with no one around: "Does it make a sound?" Von Neumann's quantum worldview is "no," because a conscious entity (to make the "observation") is required to collapse the wave functions of both the tree and the resultant sound waves.

Von Neumann makes a convincing argument. To date, no one has surfaced with any persuasive explanation of what precise mechanism causes the wave-function collapse. Although von Neumann's explanation may seem a bit metaphysical, I have yet to see a theory that is any better.

To date, scientists have already performed experiments where electrons, atoms and, recently, even large molecules behaved as waves. We are no longer dealing with theory. Solid material objects *do* exist as waves.

Contrary to what some believe, I do not believe that consciousness is the essential catalyst that collapses the wave function. Rather, my current research indicates that it is the ill-defined dimension of *time*. In the pages that follow, I will outline how this special dimension, in which we exist and waves do not, is the mystical "ingredient" that forces the wave-function collapse. I will demonstrate how the "pre-measurement" state of any wave-object (which identifies the *wave* component of the wave-particle duality of matter)—just like any electromagnetic (EM) wave—is pure energy and exists *outside* of time.

The act of observation or measurement does not directly trigger the wave-function collapse, although it is easy to understand why this perception occurred. The observation occurs simultaneous to time's *capture and transformation* of the wave to particle.

This simple embellishment of the Copenhagen Interpretation also eliminates the confusion of identifying the *type* of consciousness (human, animal, divine) that is necessary to trigger the wave-function collapse. Some experiments have even implied that light or lasers serve the same capacity as von Neumann's consciousness in collapsing the wave function. Others have further sought to invoke the divine and suggest that it is God's consciousness that explains the Universe's very existence.

Although I am a deeply spiritual person, I believe that *time* is the primary catalyst in the measurement process that triggers the wave-function collapse.

Since time and matter appear so intimately connected, I will frequently refer to this intimate relationship as the *time-matter continuum*.

Chapter 2: Einstein's Tree

History identifies that even Einstein never fully bought in to all the concepts of quantum physics, including some components of the Copenhagen Interpretation which he helped to create. He, in fact, spent many years of his life trying to disprove some elements, including the effect that observation appeared to have upon our particulate reality. His famous comments regarding "God does not play dice with the Universe," and "Do you really think the moon isn't there if you aren't looking at it?" exemplify his stance.

Following is a classic excerpt from an interview that Einstein had with fellow physicist Ernest Sternglass:

"You see the large tree over there,"

[Einstein] said. "Now turn your head away. Is it still there?"...He was explaining to me one of the principal aspects of the Copenhagen Interpretation of quantum theory that he found particularly unacceptable, according to which an observation or measurement is necessary to bring an object like an electron into definite existence.[8]

Because of this skepticism, Einstein and two of his colleagues, Boris Podolsky and Nathan Rosen designed the famous **EPR** thought experiment (1935) meant to demonstrate several presumed flaws in the current state of quantum mechanics and its Copenhagen Interpretation. (Einstein preferred to state that the interpretation was "incomplete.")

However, when the thought experiment was finally put to the test by mathematician John S. Bell (1964) and then physically performed by physicists John Clauser (1972) and later Alain Aspect (1982), the clear winner was quantum physics, not Einstein.

As noted, von Neumann singled out

consciousness as the particular determinant that collapses the nebulous, "natural" world of waves to the particulate world that we recognize. Others simply state that it must be another unidentified mechanism in the measurement process that is involved.

Either way, let us look at how von Neumann's theory might be expected to work. Let us use Einstein's tree as the example. Let us also assume that sight and touch are the two main modalities (senses) by which we will "experience" the presence of the tree.

If we neither touch nor look at the tree, the Copenhagen Interpretation states that the tree still exists, but it exists only as an obscure aggregate of tree-waves, with some waves possibly extending out an infinite distance (to be discussed) from the tree's presumed (last observed) location. It is only when we open our eyes and look in the direction of the tree, or physically interact with it (e.g., touch), that the tree-waves collapse or consolidate into the solid, particulate body that we recognize as a tree.

Our eyes see and our hands may now feel the tangible structure of the tree's trunk, bark, branches, and leaves. Let us examine both sensory modalities—sight and touch—in more detail, as von Neumann might have explained them.

Had we kept our eyes closed, imagine that we searched blindly for the tree with our arms extended. As we probed, at some point our fingertips would encounter one or more tree-waves—perhaps branch-waves, leave-waves, or trunk-waves. As soon as our hands intercepted or "touched" the outermost first wave (or waves), the corresponding collection of tree waves would collapse. We could now grasp the leaves and feel the tree's rough bark.

Similarly, had we—with eyes now open—turned our heads to look in the direction of the tree, our eyes would have somehow interacted with the tree-waves and collapsed the tree's overall wave function.

One problem that enters this latter scenario is that the retinas of our eyes respond only to visible light—light which is either emitted or reflected from observable objects. It doesn't make sense to me that our eyes would "sense" the tree's *material* waves directly. Our eyes have evolved to detect only visible light waves.

Somehow though, visible light, reflecting from the tree, acts as an intermediary to collapse the tree's overall wave function. I will discuss this mechanism in more detail in **Chapter 9**, in association with the concept of entanglement.

Von Neumann's involvement of consciousness merely adds an unnecessary level of complexity to the theoretical wave-collapse mechanism. I believe that we can safely omit this element from the equation.

Although it may seem ludicrous to think of things like trees or bowling balls as waves, this is precisely what quantum physics tells us—that macroscopic objects, under the proper circumstances, are capable of displaying qualities similar to that of the light wave! Imagine firing a bowling ball at a solid wall with two slits (as in **Figure 1.1**). The Copenhagen Interpretation states that it is *possible* (regardless of how unlikely) that the bowling ball will display its wave properties and split into two half waves before coalescing and crashing into the target.

Before continuing, we need to clarify and contrast the presumed wave functions of quantum (i.e., atomic and subatomic) particles against those of macroscopic objects like Einstein's tree or the bowling ball.

One pioneer in this arena was French physicist, Louis de Broglie. In 1924, de Broglie published a landmark paper dealing with electron waves and the wave-particle duality of matter. Although scientists have long acknowledged the wave-particle conundrum—and it has been detailed by the likes of Max Planck and Albert Einstein, de Broglie's research concluded that *every* material object has an associated wave or waves. For this research, de Broglie won the Nobel Prize in physics in 1929.

According to de Broglie, every material object may exist and behave as a wave—just as an electron or photon might. The major factor which determines our limited human ability to detect and measure each object's wave, however, is the object's *size*.

De Broglie boiled his theory down to a

single equation, not unlike Einstein's famous $E = mc^2$. For our purposes, the de Broglie equation can be summarized as follows:

$\lambda = h/mv$
    where λ (lambda) = length
       of the object's wave
    h = Planck's constant (= 6.626 x
       10^{-34} joule·second)
    m = mass of the object
    v = velocity of the object

What you can deduce from the above formula is that the larger the mass of the object, the smaller (shorter) its wavelength. Hence, a very small electron would have a much larger relative wavelength than, say, a much larger object like Einstein's tree.

Similarly, the same relationship applies to the object's velocity: the greater its velocity, the shorter its wavelength.

For the electron, for instance, we would calculate its wavelength to be on the order of about 10^{-12} meters—small but measurable.

In contrast, when we perform the same calculations for many macroscopic objects, their wavelengths turn out to be about 10^{-34} meters—orders of magnitude smaller than the size of even an electron's wavelength. This is the prominent argument by scientists for why we will rarely ever measure macroscopic objects as waves—their wavelengths are just too small.

Researchers, to their credit, have recently succeeded in measuring the wave properties of very large molecules. To measure these waves, researchers must apply thermal energy to place the target molecules in motion. One problem arises, however, when high velocities are imparted to these molecules. High velocities decrease the size of the molecules' wavelengths and increase the difficulty of measuring them. Hence, researchers heat them to the lowest temperatures possible (usually well over a 1000°F) to achieve the necessary results.

Yet, regardless of any wave's length, certain physics principles hamper the already-difficult measurement process. When any *moving* wave (such as an EM wave) travels out into space, its intensity decreases (varies *inversely*) as the square of its distance from the source. As an example, if an observer *doubles* his distance from a star, the star's brilliance will diminish to a *quarter* of its previous brightness. This relationship follows the inverse-square law and applies to any wave, like gravity or light.

When you consider the inverse square law of waves, you have to revel in the incredible energy of the photon each time you use your cell phone and witness communicating with a cell tower several miles away. The photon is striking in its strength. [We may look with the same amazement—and perhaps caution—upon our own brains—emitting electromagnetic waves identical to the waves of an electroencephalogram and on a continual basis.[9]]

In contrast to the inverse square law of a wave's intensity, the magnitude or size of a wave's expanding *wavefront* will increase *directly* (not inversely) as the square of its distance from the source, becoming progressively larger while its strength gets progressively smaller. Studies reveal that this is not the case for electromagnetic waves, however, as their wavefronts protrude perpendicular to the wave's direction of travel (**Figure 2.1**).

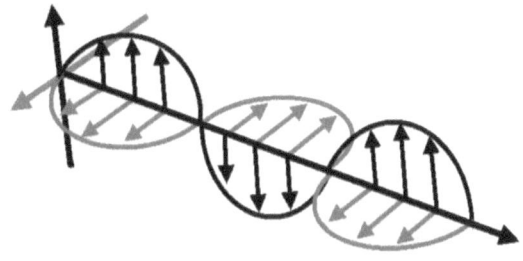

Figure 2.1 A 3-D representation of an EM wave with its magnetic and electrical components perpendicular to its direction of travel

We have to plead a large degree of ignorance when it comes to defining standing (also called *quantum mechanical*) waves. Researchers know very little about the waves of macroscopic objects since they are so extremely small and difficult to measure.

Yet, as our technologies continue to advance, scientists are measuring the waves of larger and larger molecules—once thought to be impossible.

The current version of the Copenhagen

Interpretation defines any *unobserved* wave function as possessing an infinite number of possible values. In other words, until the wave function is actually measured, it has the *potential* (even if infinitely small) of finding itself in a location on the opposite side of the Universe. Even the wave function of a material object literally exists across all space—just like the light wave.

When the defining measurement finally occurs, Heisenberg's uncertainty principle further clouds our view of reality by restricting our ability to identify *both* the object's location (position) and momentum *simultaneously*. The laws of physics allow us to identify only one—not both. Such is the limited nature of trying to pin down nature.

With such mystery surrounding the measurement process, it is not surprising that some physicists still believe that consciousness is the final catalyst of the wave-function collapse. If this is so, the question must then remain whether animals are equally capable of collapsing the enigmatic wave function. With the recent advances in technology, however, I believe we can put this question to rest.

Chapter 3: The NIST and Quantum Zeno

The NIST experiment of March 1990[10] is celebrated because it demonstrated a crucial role in how observation alters our reality. Specifically, researchers found that the experiment could not proceed properly *while being monitored*. It's as if the experiment knew it was being watched and, accordingly, refused to get dressed in front of the researchers. Predictably, one of the most popular explanations involved the Copenhagen Interpretation and emphasized that the experiment could progress only in its *unobserved*, natural wave state.

The researchers noticed that the more often they monitored the progress of the experiment, the more they delayed the final outcome. They could only conclude that their observations were forcing the wave functions within the experiment to collapse (to particles), temporarily halting the progress of the experiment.

Before explaining the details of the experiment, it is important to recall that the major point of differentiation between the wave of an electron versus Einstein's tree is the relative size of each. The scientific community has only recently considered that the *natural* tendency of matter is to exist as waves rather than particles. A short excerpt taken from the constantly updated website, Wikipedia.org, exemplifies this relatively new perspective:

> [E]lectrons do not orbit the nucleus in the sense of a planet orbiting the sun, but instead exist as standing waves.[11]

Note: this does not mean that the classic Bohr model of the atom, which most of us learned as students, is now totally obsolete. In fact, from a practical standpoint, I will employ precisely such obsolete diagrams to depict atoms in this text because that is how most of us recognize them—as electrons whirling like small dots about a central nucleus.

So, keeping in mind that electrons and atoms are most appropriately envisioned as *standing* waves (and not as dots or particles), let us continue.

In the NIST experiment, the researchers employed radio waves to excite the electrons of beryllium atoms (**Figure 3.1**).

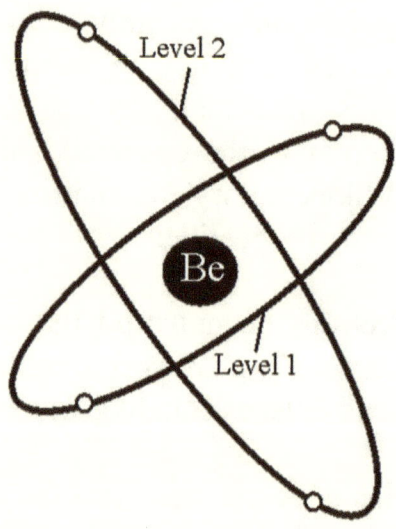

Figure 3.1 A beryllium atom

The radio impulses stimulated the ground-state electrons (in orbital level 1) to jump to the next higher orbital level 2 (**Figure 3.2**).

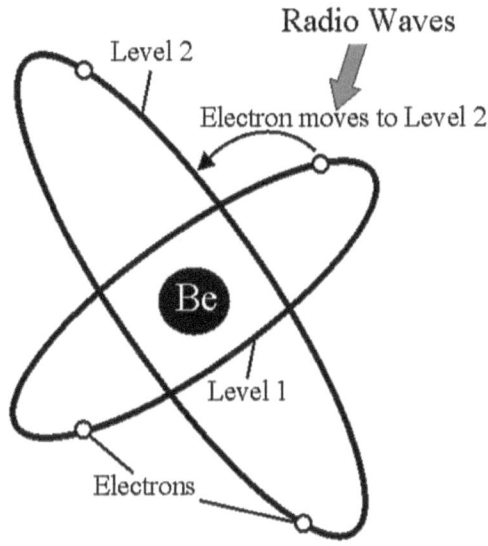

Figure 3.2 Radio waves force the level 1 electrons to jump to level 2.

The investigators found, by stimulating the electrons for a duration of 256 milliseconds, that 100 percent of the level 1 electrons shifted to the higher energy level 2 state. Likewise, an impulse of half that duration, 128 milliseconds, caused only 50 percent of the electrons to make the jump, and so on.

Researchers measured the electron transitions by employing a *laser* that measured the number of electrons remaining in level 1. This methodology allowed the team to monitor the electrons without altering any elements or processes in the experiment...or so they thought.

What the experimenters expected under classical Newtonian physics was that the measurements would show that 100 percent of the electrons would be in the starting level 1 at 0 milliseconds, 50 percent after 128 milliseconds of stimulation, and 0 percent in level 1 after 256 milliseconds (with 100% now in level 2). Once again, classical physics proved itself obsolete.

The researchers found that the number of observations directly impacted the progress of the experiment (i.e., the number of electrons making the transition from level 1 to level 2). For instance, if the examiners checked the electron states four different times during the experiment (e.g., at 64, 128, 192, and 256 milliseconds), an unexpected two-thirds of the ions still remained in level 1 at 256 milliseconds instead of the predicted 0%.

If the investigators further increased their surveillance and checked the status of the electrons as often as every 4 milliseconds (for a total of 64 times), almost *all* the ions remained in level 1 at 256 milliseconds. The researchers could essentially prevent the experiment from ever progressing—through observation.

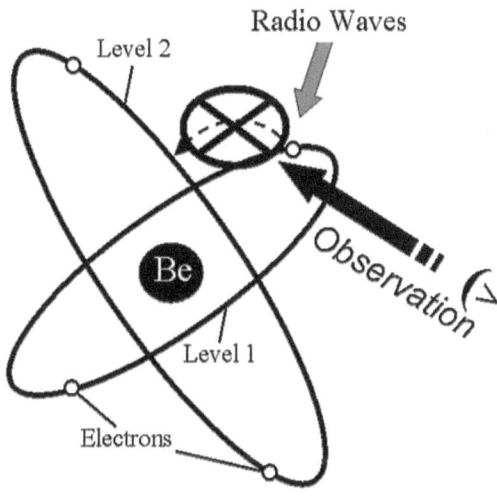

Figure 3.3 Monitoring the experiment prevented the electrons from moving to level 2.

The act of monitoring was somehow interfering with the electrons' abilities to transition to level 2 (**Figure 3.3**).

Utilizing classical Newtonian logic, the

monitoring laser pulses should not have interfered in this experiment. It is apparent, however, that *something* did interact with one or more components in the investigation and alter (i.e., delay) the anticipated outcome.

The current Copenhagen Interpretation would indicate that the sole act of measuring was enough to impede the progress of the experiment. But how?

I will reiterate that one of the most accepted explanations is that Einstein's unwatched tree exists in an ill-defined wave form until the tree is observed, whereupon it immediately collapses to its particulate, well-defined solid state. If this same concept is applied to the NIST experiment, then the act of monitoring is collapsing some critical wave component in the experiment—the literal quantum equivalent to "a watched pot never boils":

> The idea, which is a natural corollary to the idea that an unwatched quantum entity does not exist as a 'particle' [but as a wave], had been around since the late 1970s. A watched quantum pot, theory says, never boils. And

experiments…bear this out [author and physicist, John Gribbin].[12]

The *quantum Zeno effect* is the scientific term applied to the phenomenon that quantum particles, if continuously observed, will never decay. Hence, it is fitting that the same terminology is also applied to this experiment since the same mechanism is undoubtedly involved.

At this point, I should emphasize an important, proven fact regarding the EM wave: time does not exist for any EM wave. Said another way, time stops for any wave traveling at the speed of light (c). If an EM wave's clock never ticks, it can theoretically travel throughout the Universe and exist everywhere at once. Most physicists accept this premise, as unbelievable as it may sound, as true. Light waves are omnipresent—existing everywhere throughout the Universe at once!

At fractional or less-than-light speeds, time will slow (known as *time dilation*) but not stop. Scientists believe that neither material objects nor any particles of mass may travel at the *speed of light* (whereupon time would stop). They may travel at speeds less than c, however.

Consider the apparent paradox that, as a wave, a photon is omnipresent. Yet, as soon as the same photon is measured (and transforms to a particle), its speed is reduced to the finite, observed speed of 186,00 miles per second (or 300,000 km/sec). There exists a definite discrepancy here. One moment, the photon (as a wave) has an infinite velocity (with its clock stopped and existing everywhere at once). At another instant, the same photon is *measured* (as a particle) as traveling at the very finite speed of 186,000 mi/sec. The difference in apparent velocities is whether the photon has been measured or not.

Recall that the current Copenhagen Interpretation suggests that our material world exists only because it *is* observed. Such a statement begs the obvious question, "What is our reality like when it is not being observed?" Is it possible that the unobserved world and its material waves exist outside of time just like EM waves? I believe it is certain.

If all unmonitored "material" waves

(i.e., the NIST electrons or Einstein's tree) share the same traits as EM waves, then all unwatched material waves, like light, exist outside of time, are extra-dimensional, and are omnipresent.

Following are some inferences that we might draw from our current knowledge of waves and recent advances in quantum mechanics:

1. *Unobserved EM* waves are timeless, existing everywhere at once. Hence, they must, from our human perspective, have infinite velocities.

2. *Observed*, collapsed *EM* particles (i.e., existing within our dimension of time) have the measured finite velocity of 186,000 mi/sec, unlike their timeless wave-counterparts.

3. *Unobserved*, non-EM *material waves* (e.g., Einstein's tree) likely compare to their EM counterparts by also existing outside of time. In their unobserved states, these material, standing waves exist solely as Einstein's E or pure

energy—massless, timeless, and with infinitely broad wavefronts. In fact, as we shall see, the Higgs field theory (to be discussed in **Chapter 11**) presumes that all material *waves* are just that: massless—just like the unobserved light wave—before encountering the hypothetical Higgs field (which I equate to the dimension of time).

4. *Observed material particles*, however, cannot travel at light's *measured* speed of 186,000 mi/sec, but this does not prohibit their *waves* from being timeless. Any velocity (even my slow walking speed) becomes infinite in a realm devoid of time.

   Science so far agrees with Einstein that no material object may travel at 186,000 mi/sec, the speed barrier for any object ensnared within the dimension of time.[13] However, material objects may travel at fractional light speeds, and their clocks will slow. One example includes the unstable subatomic particle known as the *muon*. Cosmic rays strike molecules in the Earth's upper atmosphere to create these particles. The collisions propel the newly created particles to velocities approaching c. Measuring devices have subsequently recorded that the normally short life spans of these muons appreciably increase from the resulting time dilation of their increased velocities. Further, scientists observe that clocks aboard rapidly moving aircraft slow down relative to their stationary counterparts on Earth. Time dilation is accepted fact.

5. Separately, one other condition may slow the clocks of material objects and particles. In his general theory, Einstein revealed that *gravity alone slows time* without having to invoke the speed of light. In his general theory of relativity, Einstein revealed that any accelerating force mimics the effect of

gravity. For instance, if a space traveler accelerates at the rate of 32.2 ft/sec^2, the resultant force exerted upon the passenger is equivalent to the gravitational force of standing on the Earth's surface. Under a more extreme condition, if the same traveler goes into orbit around a black hole at the distance of its Schwarzschild radius (the boundary beyond which even light cannot escape), time will also stop—just as if you were traveling as a light wave at c. Hence, gravitational waves (which also travel at c, by the way), inherently possess a separate mechanism by which they may influence time.

As Einstein so aptly noted, "Time and space are modes by which we think and not conditions in which we live."

Science, then, concludes that both gravity and extreme speed are separate (but related) mechanisms which may slow time.[14] In addition, we must also consider that *all* waves (like all EM waves) are omnipresent (i.e., timeless) prior to observation and that all particles (even photons) assume finite speeds *after* observation. I should point out that quantum physicists are still struggling with gravity's unique role.

In summary, the act of measurement somehow negates the timelessness of any wave and transforms it to a visible, measurable particle. Turn your head away (or your measuring device), and the particle appears to return to its ill-defined, timeless state.

Likewise, approach c or a black hole's event horizon (as a material object), and your clock will slow. In contrasting fashion, assume zero velocity or a weightless environment, and your clock will speed up.

The idea that material particles (like electrons) can behave as waves when unobserved is not new. For instance, most scientists accept that, in the NIST experiment, the act of observation prevented the electrons from changing energy levels. The electrons could change orbital levels only in their unobserved "natural" states—as waves. Thus, each time the researchers peeked into the experiment, the electron waves collapsed to their particulate states, and they could not change orbital levels. When the monitoring ceased, if only temporarily, the electrons returned to their natural wave states and could again resume their shifts to the higher energy levels. As such, the more frequently the experimenters monitored the electrons, the more they delayed the progress of the experiment. Note physicist and author John Gribbin's observation:

> [I]f it were possible to monitor the ions all the time then none of them would ever change. If, as quantum theory suggests, the world only exists because it is being observed, then it is also true that the world only changes because it is not being observed all the time.[15]

Chapter 4: The Photon— Out of Sight, Out of Time

When Einstein published his special theory of relativity in 1905, the field of physics changed forever.

The genius of Sir Isaac Newton would remain, but a new king had been crowned. According to Einstein's theory, substantiated by subsequent experiments and observations, light alone remained the only constant in nature. Scientists now perceived the dimensions of space and time as malleable dimensions—no longer the fixed parameters that we had always imagined.

Prior to Einstein, the light wave, like a speeding bullet, was believed to abide by classic Newtonian principles.

For instance, consider a bullet fired by a

rifle, and it approaches me at a known speed of 1000 feet/second (**Figure 4.1**). Secondly, assume that I, as the observer, am traveling *toward* the bullet at 500 feet/second. Classic Newtonian physics predicts, and experiments verify, that the oncoming bullet, in relation to me as an observer, will be *measured* as traveling toward me at an apparent, increased speed of 1000 + 500, or 1500, feet/second. The bullet's *apparent* speed is additive to that of my own speed as an observer.

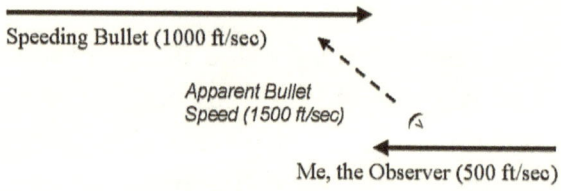

Figure 4.1 Moving observer measuring an approaching bullet's speed

Similarly, if I change my direction and decide to travel in the same direction as the bullet, the bullet would now appear to travel slower than its actual speed (of 1000 ft/sec) or 1000 − 500, or 500, feet/second (**Figure 4.2**). The math is pure Newtonian or classical physics. Prior to Einstein, scientists predicted the light wave would behave no differently.

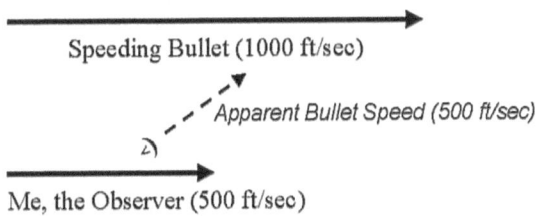

Speeding Bullet (1000 ft/sec)

Apparent Bullet Speed (500 ft/sec)

Me, the Observer (500 ft/sec)

Figure 4.2 Moving observer measuring a bullet's speed, same direction of travel

Simply put, if I (as the observer) were to change my present speed or direction of travel, then the relative observed speed of any measured, moving particle (like a bullet) or wave (like a sound wave) would also be *expected* to change—according to classical physics.

Einstein proved, however, that this was not the case for the enigmatic photon. With light (or any EM wave), experiments carried out in the late 1800s suggested that light might be acting differently—traveling at a constant velocity (the c in Einstein's $E = mc^2$) regardless of the observer's speed and direction of travel (**Figures 4.3 – 6**).

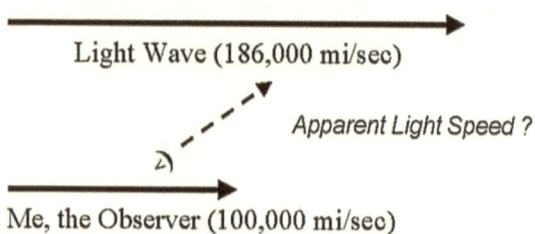

Figure 4.3 Newton's *prediction* for the relative speed of light: 186,000 – 100,000 mi/sec or 86,000 mi/sec

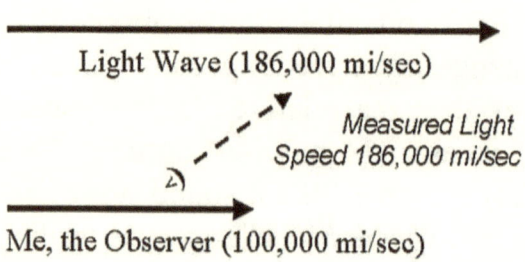

Figure 4.4 Fact: Light's speed remains constant at 186,000 mi/sec

In 1887, Albert Michelson and Edward Morley conducted an experiment to verify the existence of the mysterious "luminiferous aether," the medium through which light was believed to propagate in outer space. The results that Michelson and

Morley identified were not what they had expected.

Their experiment revealed that there was no luminiferous ether. Of historical significance, however, is that the experiment also revealed that light's speed appeared constant throughout the experiment, despite Earth's relative motion. Unfortunately for Michelson and Morley, they chalked these results up to "experimental error."

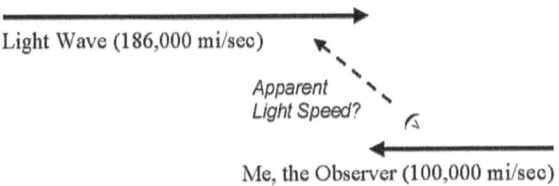

Figure 4.5 Newton's *prediction* for the relative speed of light: 186,000 + 100,000 mi/sec or 286,000 mi/sec

Needless to say, Einstein made no such error in his calculations for his special theory of relativity. Hence, Einstein is remembered for his role in proving the constancy of c, not Michelson and Morley.

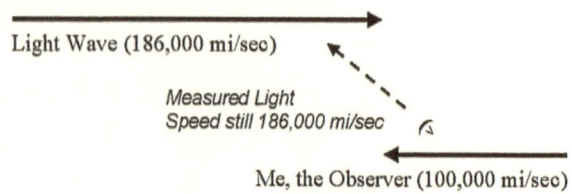

Figure 4.6 Fact: Light's speed again remains constant at 186,000 mi/sec.

Another major element of Einstein's special theory resides in the proven fact that time does not exist for the photon—or any other wave capable of traveling at the speed of light (like the gravitational wave). Or, said another way, light and other waves traveling at the speed of light exist *outside of time*. That is, these waves are extra-dimensional—existing outside of the four dimensions that ensnare us as humans.[16]

As previously noted, it is this dimensional paradox that explains why, to us (trapped within the dimension of time), light appears to travel at the finite speed of 186,000 miles/second.

Yet, for the light wave, we simultaneously *know* that neither time nor distance exists. With its clock hands frozen forever in time, light can travel any distance throughout the Universe in zero time. Light waves never age. To a light wave, distance does not exist. Light is omnipresent throughout the Universe. In fact, since time does not exist for the photon, there also exists no difference between the past, present, and future.

Author and astrophysicist John Gribbin reiterates this sentiment:

[T]ime does not exist for an electromagnetic wave, so that it is everywhere along its path (everywhere in the universe) at once; or you can say that distance does not exist for an electromagnetic wave.[17]

Everything in the universe, past, present, and future, is connected to everything else, by a web of electromagnetic radiation that "sees" everything at once.[18]

Let us consider one other ramification of light's timelessness. How knowledgeable would you be if you could be everywhere in the Universe at once—and for the past and future as well as the present? For many, the word *omniscient* comes to mind. Acknowledging such supernatural-like elements for light, many physicists literally describe the photon in divine terms.

A striking series of physics experiments exemplify—in real time—the timeless nature of light and other forms of EM radiation. The *quantum eraser* experiments were a natural corollary of the complex and somewhat confusing experiment known as the double-slit experiment (**Figure 1.1**). The double-slit experiment first revealed the mysterious, schizophrenic nature of light (and also electrons, for that matter). The outcome was so baffling that physicists and other researchers endlessly sought to clarify its meaning. The quantum eraser series of experiments revealed the same quantum mysteries of light, but in a manner that is much simpler to explain—and, hopefully, to understand.

In very elementary terms, I will explain two of the more common experimental set-ups for two separate versions of the quantum eraser experiment. I will simply call them experiments 1 and 2.

In quantum eraser experiment 1, researchers shot a single photon into the experimental set-up (the beginning). In the middle of the set-up the photon splits into two smaller photons. At the end, a measuring device records the final path of each split photon. **Figure 4.7** represents the unmodified experiment, and the researchers recorded the results.

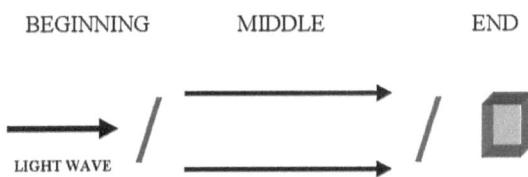

BEGINNING   MIDDLE   END

LIGHT WAVE

Figure 4.7 Quantum eraser experiment 1

In **Figure 4.8**, a modification was made at the very end of the experimental set-up, and the researchers repeated the experiment.

What the researchers observed, to their amazement, was that altering the very end of the experiment caused the two photons to alter their pathways in the *middle* of the experiment (**Figure 4.9**)! Certainly, this is not what the investigators expected, and, for many reasons, seemed impossible. What could explain these unusual results? Had they made a mistake?

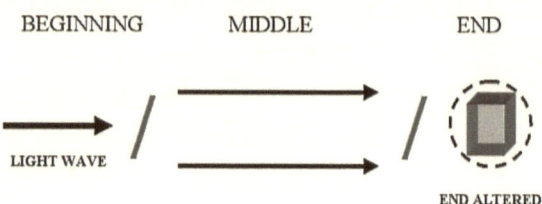

Figure 4.8 Quantum eraser experiment 1 —end is modified

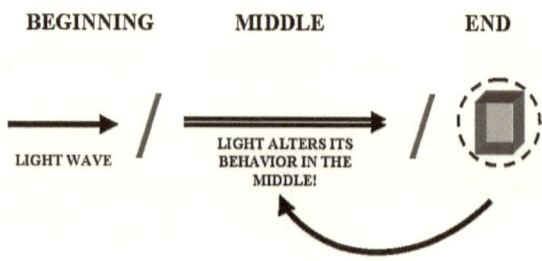

Figure 4.9 Quantum eraser experiment 1: modifying the end altered light's behavior in the middle.

According to Newtonian physics, making any change(s) at the end of any experiment (think of chemistry class) would not affect what you had already done. Yet, that appeared to be precisely what was occurring in this experiment.

I often draw the following analogy with this experiment. Consider that you are pouring water from bucket A (in your hands, the beginning of the experiment) to bucket B (sitting on the floor, the end of the experiment). Someone moves bucket B to a site five feet away, but you don't change anything you're doing. You keep on pouring the water out of bucket A just as you had been. You expect the water to hit the floor. Remarkably, however, the stream of water alters its course in mid-air and travels to bucket B five feet away. Impossible!

Before drawing any conclusions, let us now look at its companion, experiment 2.

In quantum eraser experiment 2 (**Figure 4.10**), the set-up is different from experiment 1 in that the upper and lower arms of the set-up are completely changed and are now the main focus of the researchers. Bear with me as I explain.

Similar to the beginning of experiment 1, a single photon is released into the experiment. The photon similarly splits, with one split photon proceeding into the upper arm, and the other advancing into the lower arm of the experimental set-up. The results are recorded.

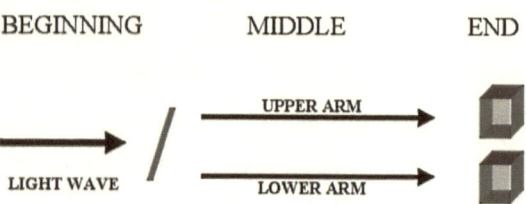

Figure 4.10 Quantum eraser experiment 2

In the modified version of experiment 2, the researchers made a single change (**Figure 4.11**) to only the *upper arm* of the set-up (not at the end, like experiment 1).

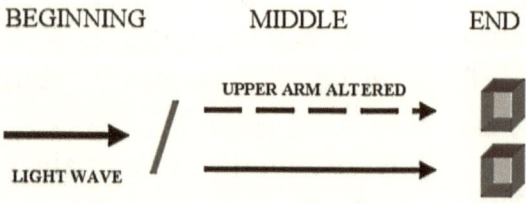

Figure 4.11 Quantum eraser experiment 2: upper arm only is modified

Yet, when the modified experiment was run, its results were no less startling than those of experiment 1. Although the two arms of the experiment were not linked beyond the point at which the photon split (in the beginning), altering the progeny photon in the upper arm (middle of the experiment) simultaneously altered the behavior of its twin in the lower arm (**Figure 4.12**)!

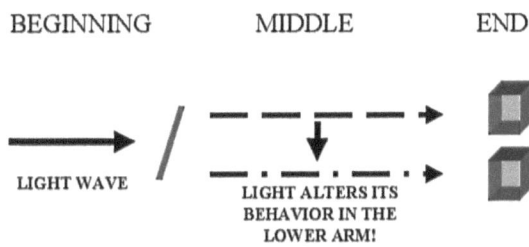

BEGINNING   MIDDLE   END

LIGHT WAVE

LIGHT ALTERS ITS
BEHAVIOR IN THE
LOWER ARM!

Figure 4.12 Quantum eraser experiment 2—modifying the upper arm alters light's behavior in the lower arm.

In each experiment, the divided photons clearly appeared to communicate—either with their past selves (experiment 1) or simultaneously with each other (experiment 2).

In experiment 1, the photons appeared to communicate retrograde, backward in time, from the end of the experiment to the middle, where they altered their pathways. The experiment suggested that the researchers were witnessing light's ability to transition between past, present, and future—outside of time.

In experiment 2, the two separated photons appeared to communicate and react instantaneously. Altering one photon immediately caused the other to change its behavior.

These experiments not only suggested that photons are capable of communication, but also, as some physicists phrased it, "consciousness."

These traits of the unique but ubiquitous photon have astounded even the brightest of modern-day scientists. Below, I quote two well-respected physicists and Nobel Prize winners on this topic:

> Those who are not shocked when they first come across quantum theory cannot possibly have understood it [Niels Bohr].[19]

> [Light's behavior is] a phenomenon which is impossible, *absolutely* impossible, to explain in any classical

way, and which has in it the heart of quantum mechanics. In reality, it contains the *only* mystery...the basic peculiarities of all quantum mechanics [Richard Feynman].[20]

As I like to tell my audiences at the end of my lectures, if you are currently shocked, confused, or in disbelief, then you probably understand quantum physics better than your peers.

Chapter 5: Entangle My Universe

I mentioned the EPR experiment in **Chapter 2**: the experiment devised in 1935 by **E**instein and two of his colleagues (Boris **P**odolsky and Nathan **R**osen) to disprove some unintended consequences of the Copenhagen Interpretation of quantum physics. For example, their original view of quantum mechanics defined that a single system (e.g., a whole unit wave) has its own "wave function." The meaning of this statement is severalfold. If a *single* wave ever splits into two smaller waves (which is done all the time in physics experiments—including the quantum eraser experiments), the characteristics of the two resultant ("twin" or "progeny") waves will be forever locked together or "entangled" since they are products of a single system or wave function. Similarly, these progeny waves will be identical in character. These entangled waves do not have to originate from a single wave, like a photon, however. Any two waves (even if whole), that originate simultaneously from a *common single source* (like a decaying neutral pion or energized atom), are also identical, entangled twins. No matter how far apart you separate them, they still bear identical or mirror characteristics (**Figure 5.1**).

A   Source   B

Figure 5.1 EPR: two entangled progeny waves are released from a common source.

In quantum physics, Einstein did not believe (or want to believe) that any two progeny waves could ever be so entangled—either through time or space. He believed that each twin wave, like human twins, would have their own unique traits or personalities. Measuring a characteristic of one twin (whose favorite fruit is, say, the orange) should not necessarily identify the same characteristic of the other twin (who might favor strawberries). Still, this is precisely what quantum physics mandates—and researchers have subsequently proven. If one twin likes oranges as her favorite fruit, then so will her twin **(Figures 5.2 and 5.3)**.

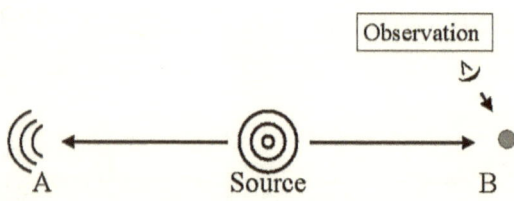

Figure 5.2 Researchers measure B's characteristics.

Einstein didn't like to think that "God played dice with the Universe" by so linking such waves. Einstein believed that such an attribute of quantum mechanics indicated a faster-than-light form of communication (or travel) between the two waves—a theoretical breach of his own special theory of relativity. This was a paradox which he, Podolsky, and Rosen hoped to clarify through their EPR thought experiment—whenever someone might ultimately be able to perform it in the laboratory. They were sure that future experiments would bear out that any two such progeny waves, separated by time and distance (i.e., not "local" in physics parlance), would have their own unique, individual characteristics.

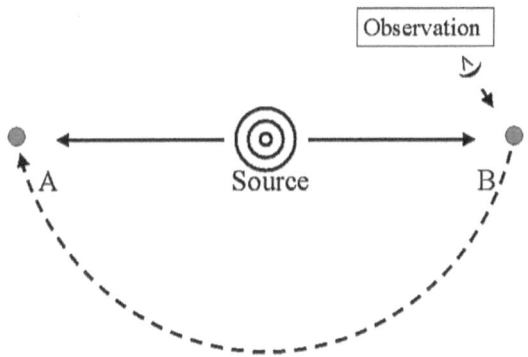

Figure 5.3 Identifying B's characteristics simultaneously forces the same properties upon A.

Yet, subsequent proofs and experiments would signal quantum mechanics and, more specifically, Einstein's own Copenhagen Interpretation as the clear victors. Two entangled waves, even when separated by space and time across the Universe, will continue to be linked by virtue of their common wave function.

Technology implied, however, that Einstein's special theory remained intact because neither a wave nor particle traveled faster than the speed of light. Rather, it was *information* (i.e., communication, even consciousness) that revealed itself to "travel" faster than light—a trait unique only to entanglement. Of course, with the acceptance of the omnipresent quality of timeless waves, researchers now have an alternative and complementary explanation.

Hence, entangled communication, one of the more mysterious attributes of waves, has proven to be a very special animal. This conscious-like behavior is a unique and separate attribute of only waves—in keeping with their omnipresence.

Scientists acknowledge that the act of measuring wave B in the EPR experiment caused the immediate collapse of the entire wave function, including that of its identical twin, wave A. The investigators had to conclude that if one wave (existing outside of time) collapsed to a particle (within time), its entangled twin had to do likewise.

This type of entanglement is the exact mechanism that occurred in the quantum eraser series of experiments and explains the enigmatic communication between the two progeny photons. This entanglement-explanation is certainly the best and most rational that science has yet to offer for the photons' actions and apparent communications.

Yet, this chapter is not complete. There is a secondary conclusion to the EPR experiment which I have not mentioned. Not only does a measurement of one wave trigger a collapse of both entangled waves to particles, but the researchers discovered another consequence—this time affecting the *observers*.

In his special theory, Einstein proved that traveling at fractional light-speeds causes multiple changes in the observer other than just the slowing of her clock. Specifically, her spatial dimensions shrink, and her mass increases. When you deal with relativity and travel at fractional light speeds, it's the *observer* who changes—not light!

With the EPR experiment, there is a similar consequence. The researchers found that if they altered the *orientation* of the measuring device at B (where wave B was measured), this action altered the reading by A's measuring device! According to classic Newtonian physics, altering the measuring device at B should only alter the reading at B. Clearly, it should have no effect at A. Yet, it does.

I compare this puzzling action to the "butterfly effect" of quantum physics: an action—no matter how seemingly insignificant—has unknown consequences elsewhere in the Universe. I even extend the definition of *action* to include our human thoughts, which are, after all, none other than EM waves that extend out into space. The results of quantum mechanics can certainly give us all extra pause—even to what we think.

Once again, we are all witness to how examining the intimate and shrouded domains of quantum physics can alter not only our own views of reality, but those of our Universe.

Chapter 6: Living in a Nonmaterial World

So, what are we to believe? Our minds and logic tell us that the world in which we live is real. It's concrete. It's particulate. We can see it, touch it, taste it, smell it, and hear it.

Science, on the other hand, suggests something else. Our minds may be deceiving us. The world, when we are not observing it, is totally different. We have always believed, when we closed our eyes or slept, that the world maintained its basic shape and substance...that is, things would be essentially the same when we woke up as when we went to sleep.

The Copenhagen Interpretation tells us, however, that this simple assumption is not necessarily the case. When the world sleeps, our Universe may transform into a nonmaterial fairyland of ghostly waves and electromagnetic energy. Nothing is real—as we know it—anymore. Yet, when we awake, everything is just as we left it—or is it?

Are my car keys really where I left them last night?

How is it that my missing reading glasses suddenly materialized in a drawer that I *know* I checked multiple times?

I measured that wooden board three times—how is it that it is now an inch short?

The Copenhagen Interpretation may have the answer to these age-old puzzles. Specifically, it states that, up until the time of observation, my car keys exist as a host of probabilities—as waves, including the possibility that they are now two inches, two feet, or even two miles from their last observed location.

Specifically, a wave can spread out over an infinitely broad region (**Figure 2.1**). Therefore, it does not have a specific location and can theoretically be anywhere in the Universe—just like a light wave.

Statistically, however, under the prospects of possibilities, the odds are close to 100% that the keys will re-collapse in the exact location that they were the day before. Yet, there still exists some probability that their waves might collapse in a different location.

Could the Copenhagen Interpretation explain the common-day aggravation of not being able to locate your keys? Perhaps. It's an intriguing thought and certainly something that any good spouse should keep in mind.

I propose, however, that our day-to-day reality is exactly as we experience it because we are trapped within the transformative dimension of time: the time-matter continuum. If we, like the photon, could exist outside of time, things would appear quite strange.

If we could hitch a ride on the back of a photon, our past realities would exist no differently than those of the present—or even the future. We would become omnipresent and omniscient. We would have all the answers to the Universe.

Imagine also how the Universe might look. Like a photon with a frozen clock, we could literally "see" and experience the vast limits of the Universe at one time. My mind can not begin to grasp at such a reality. Nearly all of our human senses as we know them would become obsolete. Our senses would be inadequate to deal with the torrent of information flooding our minds.[21]

We would exist as pure energy and consciousness—whatever that may be like. Many experiencers of near-death have observed that, during their time "on the other side," they had all the answers to the Universe, just like the all-knowing light wave. Yet, despite this omniscience, experiencers of near-death continued to feel the very human emotions of love, regret, shame, and forgiveness. Within this new realm, they could still love and were extended love in return—the ultimate display of divine energy and Light.

Former material barriers would no longer restrict our spirits. Our Light energies could probe the deepest chasms and tiniest recesses of the material world, including the most impregnable solid materials. Our spirit energies might likewise stimulate an atom's electron to change orbital levels—or incite a radio to play.

If alternate Universes exist, we could possibly visit them as well. If the laws of physics varied in these other Universes, as they undoubtedly might, we would "experience" them differently from our own Universe with its own defined set of physical laws. The life forms of these parallel Universes would possibly differ dramatically—let your imaginations run wild. If you can think it, it possibly exists. As pure energy, we could experience it all.

In our current reality, our dream-states are likely the closest experience, for the average person (exclusive of clairvoyants and mediums), to such a mystical realm. During sleep, our electromagnetic brain waves travel unheeded into space and are capable of intermingling with the waves of other spirits—past, present, and future…good and evil. These small glimpses into the infinite netherworld of timelessness help to explain our wild confusion with dreams. The absence of time is very disorienting to us humans.

Although I believe many of the images and sounds in dreams may well originate from our own cerebral neurons, I also believe that many of the images and sounds emanate from the ubiquitous network of collective thoughts and energies that pervade our electromagnetic Universe.

Imagine our mind-waves interacting with two other persons' thoughts involving, for instance, an image of a bonfire and another of a family dog. The dreamer might well dream that his own dog is rescuing him from a frightening house fire. Yet, as the mind struggles to integrate these widely differing scenarios, any result could ensue. You might not recall whose house you were in, how the fire started, where Fido came from, or even what you were doing when the fire broke out. Let's face it, little makes sense in our dreams, and time-lines are nonexistent.

Dream interpreters would argue otherwise. There is supposed to be a lot of symbolism in dreams, and I used to believe that. Now, however, I find myself constantly readjusting my beliefs to fit with the rapid advances in science and technology.

Yet, dreams may be the closest that most of us can venture to visualizing this foreign, unobserved world.

Science suggests that the strange realm

of timelessness exists as pure energy—embodying the characteristics of omnipresence, omniscience, omnipotence,[22] and consciousness made manifest. The forces within this province are instantaneous and simultaneous.

The word *one* takes on a whole new relevance. According to Einstein, energy is mass, and vice versa. Past = present = future. There is no death. We are Einstein's "E," "m," and "c^2" rolled into one. There are no borders, boundaries, or limits. We are one with God and His Universe.

Is God real? I believe He is and that He exists as the purest form of all energies.

Further, if the angels and God are the spirits of literal Light that I presume, then our brain waves have the potential of interacting with Them as well. Hence, science now offers a rational explanation for near-death experiences, divine visions, and even prayer.

Indeed, the mystical science of quantum physics offers a plausible explanation for all supernatural and paranormal phenomena.

So where does that leave us? To start, we need to understand our place in nature and within the Universe. We need to realize our limitations and understand that we are vulnerable to many outside influences—good and bad. God has, for unknown reasons, ensnared us within this oft-confusing, four-dimensional world for a purpose.

I believe this purpose is several-fold: to love, to learn, and to leave this world a better place.

We might well wonder why some of us have received special gifts by which to experience the non-material, unobserved world (e.g., clairvoyants), while others (like me) haven't. I know my reason.

Some of us have the ability to converse with spirits. Some of us can read minds (though, granted, not predictably). Some of us can predict the future (even if haphazardly). Some can do many of the above—but only when sleeping. We have all heard such stories. I know I have. Many are credible, and many are not—just like the many messages and ideas that occasionally creep into our minds.

I believe the majority of exceptional thoughts that we have as humans propagate from the neurons within our own remarkable brain tissue, but I also believe that some ingenious ideas leak in from the outside. The latter may include those few-and-far-between "eureka moments" that we can't explain Unfortunately, some outside thoughts may also include those disturbing mental intrusions that make us crazy.

Such is the reality in which we live. We need to accept our humble positions in this life for what they are. We are here to love, learn, and adapt.

Some of us have immense power over others, while many of us are incredibly vulnerable.

The material world certainly has its limitations. It is difficult and challenging. Yet, there is a reason for everything. Once we learn to get it right, we will reunite with God in His perfect province—the dimensionless realm of spirit, Light, and love. Then, we too shall be omnipresent and omniscient. For this life, we will have to be content with not walking through walls.

Chapter 7: My Particulate Life

We wake up in the morning and vaguely remember a dream, but the details are gone.

Experiencers of near-death recall that they had all the answers to the Universe when they were on "the other side," but upon returning the answers have vanished.

A survivor of a miraculous rescue remembers following a Light to his only avenue of escape.

An ingeniously creative thought enters an artist's mind. She doesn't know where it came from, but she seizes it and runs with it—pure genius of an idea.

All of the above scenarios are true stories and relate to the non-material or unobserved fantasy world with which clairvoyants and mediums are well familiar, though they themselves may not fully understand the source.

The rest of us, on the other hand, are stuck with our material, day-to-day realities, which for the most part we perceive as the *only* reality.

Indeed, this observed world is real—it's just not the *natural* world from which our souls arose and to which we shall all-too-soon return.

We are frozen within a particulate, temporary realm—one which cannot compare to the non-material, omniscient province that harbors all knowledge, communication, consciousness, and every form of paranormal and spiritual activity with which humankind has struggled for ages to define.

Recall that as beings trapped within time, we view light as traveling at the finite speed of 186,000 mi/sec. Simultaneously, we *know* that neither distance nor time exists for light. The photon is everywhere at once.

We have noted that this paradox of light's finite speed (from our perspective) and light's simultaneous omnipresence (from its own perspective) is explained by our human entrapment within time. If we could free ourselves of this restriction, we would not see light as traveling. It would be everywhere at once—so impossible for us to imagine.

Yet, once any of these unmeasured waves are observed, they become particulate—transfixed by the literal Medusa of time.

When researchers measure an electron's orbital level with a laser (e.g., the NIST experiment), the act of measuring or observing propels time into the equation. It's as if the researcher literally stuck a magic wand into the experimental set-up. The experiment becomes frozen in time until the laser measurement stops. The electrons then return to their native wave states, and the experiment proceeds. In the NIST experiment, the laser light somehow transfixes the progression of the experiment as surely as Einstein's eyes transformed his tree.

With his eyes, Einstein collapsed the tree waves to their recognizable, particulate nature. Similarly, our own retinas freeze object-waves in time and catalyze the chain of events which allow our brains to "see" the objects.

How might the Medusa of time cause this to happen? No one yet knows the precise answer, but technology allows us to make an educated guess. The mechanism must be analogous to that which researchers witnessed with the quantum eraser experiments, where the photons clearly responded to changes in the experimental set-up without regard to time—but always part and parcel of an observation.

In the first quantum eraser set-up (**Figures 4.7-9**), the photons detected the experimental alteration at the very end of the set-up and traveled backward in time to alter their paths in the middle of the experiment—still as waves. If we can acknowledge that the quantum eraser light waves can travel and alter their pathways and behavior without regard to time, then we might consider that all waves can behave likewise.

Let us turn the tables and ask what it must be like for an omnipresent light wave to become ensnared within *our* four-dimensional reality.

Imagine that you are a light wave. You

are omnipresent and one with the Universe. Suddenly, a researcher, somewhere in space and time, takes a measurement of you, observes you, or halts your motion through space by placing her hand directly in front of you. You immediately collapse from wave to particle, ensnared within a strange, new dimension involving matter and the ticking of a clock.

Or imagine that you have just collapsed onto a human retina. You are no longer traveling unimpeded through space. You are now a part of the retinal membrane, ensnared by an electron—in a rhodopsin molecule specifically. The energy that you have imparted to the electron alters the chemical make-up of the entire molecule, initiating the first step in its transition to retinene and the miraculous chemical cascade that we call sight.

Under either of the above scenarios, your future is now in limbo and totally unpredictable. As an entrapped light *particle*, you must wait until the environment and laws of physics permit you to return to your natural wave state—but even that moment may not last long.

If we look at the scenario involving the photon's capture by the retina, its potential re-emergence as a wave might last only briefly until it collides with another adjacent molecule, and its wave function collapses again. This cycle would repeat—seemingly endlessly.

Once you are ensnared by matter and the surrounding envelope of time, you can be there a long time, relatively speaking. Time defines a new existence for you as a light *particle*—one with which you were unfamiliar as a wave. You will likely lead a malleable existence, continually transitioning between wave and particle until you finally break free of the bonds of the time-matter continuum.

We may ask if the photon is really entrapped within our four-dimensional reality or whether it just appears that way to us—an illusion. Is the light wave really transformed to a particle, and, if so, has its speed truly diminished (to 186,000 mi/sec) in the process?

Indeed, if at any time the omnipresent photon passes to an *observed* velocity of 186,000 mi/sec, then its existence outside of time *has ceased.* Any "observations" involving light's omnipresence involve its existence as a wave (e.g., the quantum eraser experiments), not as a particle. Time is absent for light only in its natural wave form. Whenever scientists measure the photon's speed, it is always as a particle (within the dimension of time) and at 186,000 mi/sec. Only when researchers observe light as acting retroactively in experiments (e.g., the quantum eraser series) do they actually witness waves' omnipresence (and "infinite" velocities).

Consider also that photons are defined as *massless* particles *at rest*. A photon acquires mass only by virtue of its incredible speed. The presence of mass in any *resting particle* (i.e., matter) prevents it from ever accelerating to the speed of light. There is just not enough energy to propel any particulate object to this velocity.

Further, realize that 186,000 mi/sec is defined as the speed of a measured light particle *in a vacuum*. We know that light's speed is reduced when traveling through transparent media such as water or quartz crystal (known as the refractive index of the material, n). When we measure light's speed as slowing through these media (let's look at water with a refractive index of 1.333), light slows only because each wave's length is transiently shortened while under the field effects of the water's excited electrons (to be discussed in more detail in **Chapter 10**). Within the voids between the water molecules (and its electrons), the light waves resume their previous wavelengths and travel at c. Light's total transit time through the water is delayed only through its countless interactions with the water molecules and their electrons.

Even through our Earth's atmosphere, light's speed is impeded slightly, with the refractive index of air equaling 1.000293. Only within the true vacuum of space and the expanses between molecules does light truly travel at c.

Life within a particulate world is restrictive. Don't we know it.

Chapter 8: Waves and Strings 101

Physicists use *photon* as an all-inclusive term for *electromagnetic wave.* Do not confuse these waves with *electromagnetism.* If you look up *electromagnetism* in the dictionary or Internet, you will become lost in an avalanche of confusing and complex terminology dealing with the dual nature of a force possessing both electric and magnetic field components. *Merriam-Webster* defines *electromagnetism* as:

> [The] fundamental physical force that is responsible for interactions between charged particles which occur because of their charge and for the emission and absorption of photons, that is about 100 times weaker than the

strong [nuclear] force, and that extends over *infinite* distances [my emphasis] but is dominant over atomic and molecular distances.[23]

Could anyone have asked for a more confusing definition?

For this reason, I have learned that it is best to keep *electromagnetism* separate from the definition of an electromagnetic *wave*. Although related, I will not endeavor to discuss electromagnetism, and I mention it only to urge you not to confuse it with the wave of the same name. *Electromagnetism*, as *Merriam-Webster* notes, is a force between *charged* particles. An electromagnetic wave, on the other hand, is pure energy which possesses no resting mass and has *no charge*.

An EM wave has the added attributes of polarization (its orientation in space), direction, wavelength, and frequency.

We also know that a relationship exists between the *observed* speed of the light wave (c), its wavelength, and frequency:

$$c = \lambda \, (\text{wavelength}) \cdot f \, (\text{frequency})$$

Attempt to find an illustration of any electromagnetic wave, and, typically, you find the illustrations as show in **Figures 8.1 and 8.2**. Keep in mind, any drawing cannot do adequate justice in attempting to depict a wave—mainly because we do not yet understand exactly what a wave is. In reality, we do not understand particles any better. We are still learning.

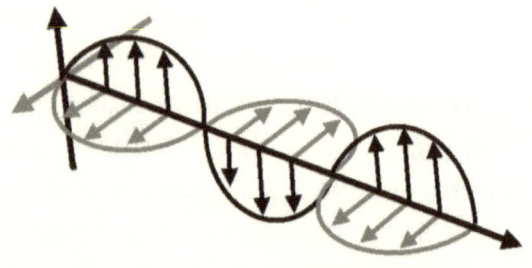

Figure 8.1 A 3-D representation of an EM wave with its perpendicular magnetic and electrical components

Figure 8.1 depicts a typical EM wave. Physicists define each EM wave as possessing an electrical and magnetic wave component or wave vector. The electrical wave vector travels at a 90° angle to its magnetic counterpart. Each wave vector has magnitude and direction. By convention, the electric field vector depicts the polarization angle of any EM wave (and its typical 2-D representation as shown in **Figure 8.2**).

Figure 8.2 Simple EM wave as commonly depicted in two dimensions

At this point, I would be negligent if I did not bring up the modern concept of string theory. As pointed out previously, the standard model of the atom that we learned in grade school is already outdated.

One attempt to replace our antiquated concept of matter and space involves several theories utilizing the moniker of "string theory." Two prominent theories go by the names of the *bosonic string* and *superstring*. What is particularly intriguing about these theories is their *requirement* for the existence of other dimensions, not just the four (including time) that we live in. For instance, the bosonic string requires a total of twenty-six (26) dimensions, and the superstring requires ten (10).

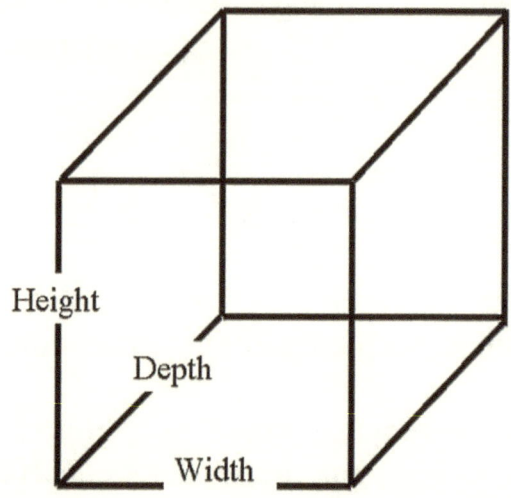

Figure 8.3 Our three spatial dimensions

Scientists like to describe these extra dimensions by comparing them to "rolled up dimensions." Try to visualize each different direction or "line" of each of our three spatial dimensions (height, width, and depth, **Figure 8.3**) as lengths of a garden hose (**Figure 8.4**). This hose, when viewed from a distance, appears to be a single dimensional line (representing, say, width). Yet, only when viewed up close can you discern its multi-dimensional nature as a 3-dimensional hose. These extra dimensions are so small that this analogy is a common way of describing the extra hidden dimensions of string theory within our limited four-dimensional existence.

The concepts supporting string theory add extra fodder to the acceptance of multiple extra dimensions. Certainly, the extra-dimensionality of light (existing outside of time) is the ultimate proof of the presence of these extra dimensions.

Figure 8.4 A single spatial dimension as represented by string theory

Light is predictably complicated and likely far too complex for our race (and certainly me) to fully comprehend any time soon. I have written entire books discussing light. Yet, in our quest for ultimate understanding, we should never abandon our search for knowledge. God wants us to ask questions. Seek, and ye might just find.

Chapter 9: Collapsing Waves...What's the Deal?

We are left with the outstanding question: what is the exact mechanism that matter and the dimension of time utilize to trigger the magical collapse of wave to particle?

To answer this question, it is best if we reconstruct what we do know. Waves are synonymous with the *absence* of time. The omission of time from the definition of a wave serves to explain most (if not all) of its mysterious characteristics.

On the other hand, particles appear to be synonymous *with* time. Similarly, mandating the presence of time for all matter explains most of its more esoteric qualities.

By observing, by measuring, and apparently just by opening our eyes, we inadvertently interact with waves, trapping them within the time-matter continuum—and halting, at least temporarily, any waveform interactions in progress (e.g., the NIST experiment). Somehow, our mere human actions, as passive as they may appear, have the ability to collapse waves to solid particles.

Whatever the exact mechanism, it is also at work when we measure light as traveling at 186,000 mi/sec—realizing that the very same light *wave* previously had no such speed limit and was literally everywhere in the Universe at once. Despite *knowing* that the photon has no finite speed limit in its "natural" wave state, we must also accept that, within *our* dimension of time, 186,000 mi/sec is precisely how we see light, experience it, and measure it.

As the NIST experiment so successfully demonstrated, the act of measurement ensnared each wave within time and collapsed each electron's wave function, halting the progress of the experiment.

Perhaps this should come as no great surprise. The act of observation or measurement *requires the element of time*.

When the local weatherman makes an observation on the local rainfall, he necessarily includes the time at which the measurement was made, knowing that the amount of rain may likely change over the next few minutes. Similarly, an optometry reading of your eyes today may not be the same in ten years. Assuredly, a photograph captures an image at a defined moment in time. You can't take a photograph of yourself from twenty years ago and try to convince someone that it was taken today. A measurement must include documentation of the time at which it was made.

The act of measuring requires the element of time: the element that transforms wave-energy into particulate, solid matter. Since we can visualize, measure, and experience waves only within our four dimensions—as particles, this is our reality. We do not ever experience waves as waves. Our reality is purely a material one—and it necessitates the element of time.

Most of us have never questioned the meaning of time. It is one of those few nouns that seems to be self-explanatory.

Yet, I thought this text would be incomplete without a formal definition. I decided to include two common definitions and one unique one.

The website Wikipedia defines *time* as "the experience that events have duration and occur in sequence with intervals between them, and also a quantity which can be measured, as by a clock."[24]

Similarly, *Merriam Webster* defines *time* as "the measured or measurable period during which an action, process, or condition exists or continues."[25]

One physics definition even defines *time* in terms of *entropy* (the concept of increasing disorder over time). *Entropy* is also known as the second law of thermodynamics. This law states that the Universe began with zero entropy (i.e., zero disorder, maximum order) and is continually progressing toward maximum entropy (maximum disorder, zero order).

Once again, I turned to *Merriam Webster* for a common definition of *entropy*:

2 a: the degradation of the matter and energy in the universe to an ultimate state of inert uniformity **b**: a process of degradation or running down or a trend to disorder

One oft-cited element of entropy includes the precept that heat always flows to regions of lower temperature. In other words, organization tends toward disorganization. A wooden board decomposes to sawdust; cars break down; homes fall into disrepair, and our bodies age (and die...and ultimately decompose). The Universe began with the Big Bang (infinite order, zero entropy) and is progressing to an empty Universe containing only photons[26] (zero order, infinite entropy).

One might reasonably argue that there exist multiple examples against the concept of entropy: for example, the construction of new buildings, the creative gestational and birthing processes, and any other process that generates design from chaos. Yet, science reveals that the energies required for these expenditures far outweigh the entropic benefits (as beautiful and amazing as the results might be).

By definition, entropy increases over

time. As such, it is easy to understand why many scholars use one to define the other.

Certainly, most people find it challenging to define *time*. Many will find themselves using *time* in the definition. Try it the next time (see what I mean?) that you're out with friends.

Some of our brightest and finest have even denied time's literal existence, including physicists and philosophers who have argued that time is merely an illusion of our human existence.

British physicist Julian Barbour, for one, maintains just such a view. He and I have a lot in common. Opting to work outside of the restrictive academic setting, Barbour focuses his attentions solely on his personal areas of interest. As such, Barbour has made the determination that the dimension of time is a human-centered delusion.

Our perception that time is real has created a host of problems in the field of physics (many of which I have already discussed). Barbour contends that, in reality, there is *no change* and *no motion*. Compare Barbour's absence of motion to my discussions of the timeless dimension where waves reside and are omnipresent. As a wave, there's no need for movement.

Barbour isn't alone. In 1982, he and fellow physicist, Bruno Bertotti, published a paper on gravity which offers solutions as accurately as Einstein's general theory of relativity—but without the requirement for time. Physicists have determined that such an approach would even obviate the need for hypothetical dark energy.

Note Barbour's view:

> In ordinary quantum mechanics, the wave function is defined on the possible configurations, which are defined in a definite inertial frame of reference, at different times.
>
> Nothing like this can happen in the wave mechanics of the world. There is neither time nor frame.[27]

To add to the growing list of supporters, many philosophers have long considered that time is a human illusion. Consider J. M. E. McTaggart and his books, *The Unreality of Time* and *The Nature of Existence*:

McTaggart concluded the world was composed of nothing but souls…McTaggart believed each of the souls (which are identified with human beings) to be immortal and defended the idea of reincarnation. *The Nature of Existence* also seeks to synthesize McTaggart's denial of the existence of time, matter etc. with their apparent existence.[28]

The perception that "time is an illusion" may paint a confusing picture. Many interpretations for explaining our perception of time exist and may include any of the following:

1) Time is a pure construct of the human mind. It is not real.
2) There exists a mechanism for the collapse of the wave function that causes a multiplicity of local—but temporary—bubbles of time. Outside of these local "realities," time does not exist.
3) A permanent subset of our Universe exists within time (where we reside). The remaining

(unobserved) Universe (dark matter/energy?) exists outside of time.

4) Our Universe is just one of many parallel universes, and ours just happens to exist *within* the dimension of time. Other multiverses exist outside of time.

Note: This list is not intended to be all-inclusive.

Since all matter naturally exists as waves (their unobserved state), we can reasonably ask whether waves have ever departed from this extra-dimensional province—that our perception is merely the illusion as hypothesized in 1 above. Unfortunately, it is not that easy.

We can infer that light waves, once they collapse onto a photographic plate for instance, no longer exist in their natural wave forms. The chemicals (and their electrons) within the glass have entrapped the photons and transformed their energies into the chemical reactions which allowed the researchers to observe *where* the photons landed on the plate (the measurement).[29]

So, what has changed about the atoms and electrons of the measurement medium, and what triggered the photons' collapse? In several experiments (like the NIST), *electrons* transitioned from waves to particles in the act of releasing *photons* (the very transformations that allowed the NIST experimenters to make their measurements of the electrons' orbital locations). Recall that the photons which allowed the NIST researchers to make their measurements radiated from the level 1 electrons.

Yet, the sole release of the photons from the electrons should not alone explain the wave-function collapse of the electrons since similar electron-photon interactions occur constantly (e.g., within various transparent materials), and the photons' wave functions remain intact. Plus, many experiments, such as in the quantum eraser series, utilize just such photon releases for the production of the single light waves that enter the experimental set-ups at the beginning. These photons clearly act as waves for the entire first half of the experiments and have *not yet collapsed*.

Still, the collapse of light waves to

particles plays a major form of *measured* wave collapse in many science experiments and, for that matter, most of our day-to-day reality. Examples of this form of observation include photons collapsing onto photographic plates, photons triggering photon detectors, electrons collapsing into dots on phosphor screens—which in turn release photons that we observe, and photons reflecting off of Einstein's tree (and all the other particulate materials of our existence) and subsequently collapse onto our retinas.

Most of what we see is the result of light reflecting off of material objects. Reflection is merely a process of photon absorption (by the object) and immediate release. When various frequencies of photons strike the leaves of Einstein's tree for instance, the green leaves absorb *all* the frequencies of light—but the green frequencies are immediately released back as reflected light. Mirrors, on the other hand, reflect most all the frequencies of light.

Most forms of sight involve wave-function collapse of photons onto our retinas after the light waves have been reflected (absorbed and then emitted) by material objects. This reflective process occurs so rapidly that we consider it to be essentially instantaneous.

The reflection of light (think of a mirror), much like its transmission through transparent media, does not appear to collapse light's wave function. The *non-reflective absorption* of light is a different story.

If the absorbing material is a photographic plate, a photon detector, a phosphor screen, a wall, or the human retina, these materials interact with light in a special way. Light *absorption* (in contrast to reflection or the transmission of light) *inherently alters the light waves*—i.e., the photons become part of the matter as a material chemical (e.g., retinal), electrical (e.g., photon detector), or thermal reaction (e.g., heating of a dark object). Hence, any type of non-reflective photon *absorption* by matter (i.e., chemical, electrical, or thermal) must exist as a major contender for explaining the mechanism that causes a wave to collapse to a particle.

We know that the collapse of one entangled twin (B) will also force the collapse of the other twin's (A's) wave function. Consider further the possibility that any wave might share its common wave function with the source that produced it—not just with its twin (should one exist). This is exactly what appears to happen in relation to the intimate association that photons have with their source electrons. For example, if a photon is entangled with the electron that emitted it (see **Figure 5.1**), and the photon is observed or measured, the source electron (as well as the photon) collapses to a particle. Several recent studies suggest that emitted photons have the capability to not only entangle their source electrons, but the source atoms as well.

Hence, the collapse of reflected photons from Einstein's tree (onto the retina) collapses also the tree's source electrons and atoms. Consider the additional possibility that not only would our material eyes collapse Einstein's tree, but so would any nearby material object (that also absorbs the tree's reflected light). This entanglement theory could further explain why it's so difficult to measure waves of macroscopic objects—there's a constant reinforcing mechanism that serves to keep matter as matter. Think about it. If Einstein's tree reflects the Sun's light, and this reflected light is then absorbed by any nearby object, then this secondary object serves to propagate the tree's state of matter (and vice versa—the tree absorbs the object's reflected light and reinforces the object's state of matter). This multiplicity of entangled waves interacting with matter serve to reinforce and maintain the local time-matter continuum.

Consider the further possibility that if a photon is entangled with its source electron and atom, then it might not be a giant leap to infer that collapse of an atom might collapse its surrounding molecules as well. A cascade of events could well follow, collapsing the entire tree. Where might the cascade stop? Might collapse of Einstein's tree function continue to the ground into which its roots extend? If so, the cascade might rationally go on to collapse the entire Earth.

As you can see, quantum physics is not the fine art that many might suppose. Various possible explanations exist for the mysterious wave-function collapse, and science cannot yet prove or disprove many of these possibilities.

It might well be that the collapse of one wave function causes adjoining wave functions to also collapse—for any number of reasons, just as one exploding firecracker might ignite the entire case of fireworks.

Yet, having considered these possibilities, even more collapse mechanisms may be at play. As contrasted to the definitive absorptive processes of light just discussed, we are about to see that there is a unique and unusual form of light *transmission* that also collapses the photon's wave function.

Chapter 10: A Polaroid of Polarization

My research indicates that the very integrity of a wave must be altered to effect its entrapment within time to form matter. The non-reflective absorption of photons by opaque materials appears to offer one such mechanism for the transformative wave-function collapse.

The ordinary transmission of light waves through a transparent medium, however, will not induce such a collapse. We know, in fact, that a light wave can pass through most transparent materials and maintain its natural wave configuration. Scientists have witnessed this in countless experiments. When a photon transits the material, its velocity decreases in proportion to its refractive index, but the photon exits the material unscathed. It appears that only when a photon impacts an opaque medium—and is *absorbed*—does it's wave function actually collapse: the measurement.

If a light wave strikes an opaque material with the proper characteristics (e.g., color), the electrons within the material will completely absorb the light wave and convert its energy into motion (heat). Take, for example, a beam of white light (consisting of all frequencies) that strikes a green leaf. The green chlorophyll molecules absorb all colors (frequencies) of light, but immediately reflect back the green. Any non-reflected light waves are absorbed, surrendering their inherent EM wave characteristics and undergoing a drastic metamorphosis. They become ensnared as material constituents of the leaf. They may never again assume their previous attributes, though they may later escape their material bonds in the form of thermal radiation.

For the reflected green frequencies coming off the leaf, as with reflected light from a mirror, the light waves maintain their intrinsic attributes and their wave functions also remain intact.

Light's transit through a transparent medium, however, is quite different to that of absorption. As light passes through a transparent medium, its velocity slows (refractive index, n, >1) compared to its travel through a pure vacuum (n = 1). All of light's inherent attributes (polarization, frequency, wavelength, etc.) will *emerge* intact—though that is not the entire case *during* its transit.

From a physics perspective, an electromagnetic wave's velocity is slowed within a transparent material because its electric and magnetic fields disturb the electrons within the medium. These excited electrons respond in kind with altered fields of their own, acting to shorten the photons' wavelengths and thereby slowing the light's velocity. As long as the light remains within the medium (stimulating the electrons), its velocity will suffer by virtue of a reduced wavelength.

Another way of remembering this effect is to recall the formula $c = \lambda \cdot f$. Since we know that a transmitting medium slows light's speed (represented by c), either its frequency (f), wavelength (λ), or both must decrease. In this case, it's light's wavelength (λ) that's affected.

Upon emerging from the medium, the light waves no longer encounter the extraneous influences of the medium's electrons, and the photons will assume their previous attributes (including their former wavelengths and speed).

It appears that for the wave collapse of a photon to occur, its wave's characteristics must inherently change in some critical fashion. Processes involving chemical, electrical, or thermal reactions within opaque media appear to meet this criterion.

For completeness, however, I must cite one additional process—one involving a specific type of light *transmission*—one which is not all it appears to be. This particular form of light transmission appears to inherently alter one important attribute of light—important enough to catalyze its wave-function collapse. What I am referring to is the process of altering the characteristic of light known as its *polarization* (its orientation in space)—and in a very *specific* way.

If you refer back to **Figure 8.2**, you can

visualize how photons assume a defined orientation (e.g., horizontal or vertical) in space. By convention, the polarization angle is defined in terms of the photon's electric field component (**Figure 8.1**), keeping in mind that the magnetic field is identical but perpendicular to the electric field.

Researchers know that by passing unpolarized light (made up of light of *all* polarization angles) through a polarization filter (polarized sunglasses, for instance), a little less than half of all the incoming light passes through the glass. The remainder of the light is either reflected or absorbed (**Figure 10.1**). This is the principle behind the glare-reducing capacity of polarized sunglasses.

Empirically, we can deduce that the wave functions of any *reflected* and *transmitted* photons do not collapse—they maintain their wave configurations.

Typically, light waves whose polarization angles are within 45° of the filtering angle (or polarization axis) of the glass will also pass through. This principle explains why nearly half of the incoming light is transmitted through a polarizing medium (**Figure 10.2**).

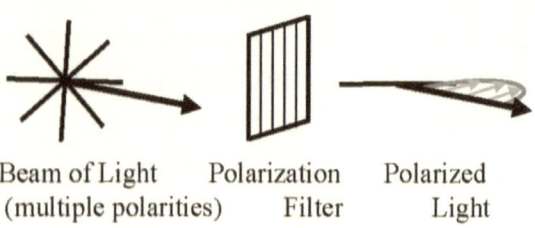

Beam of Light   Polarization   Polarized
(multiple polarities)   Filter   Light

Figure   10.1   Polarization   filter —illustrating the passage of one wave with its electric field component

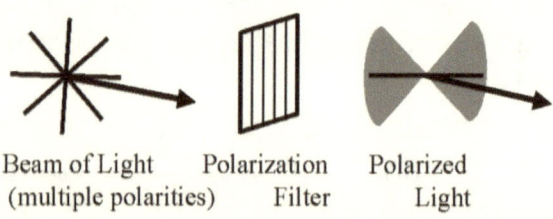

Beam of Light   Polarization   Polarized
(multiple polarities)   Filter   Light

Figure 10.2 Polarizer transmits nearly 50% of the light

The polarization filters discussed above are the most common type of polarizers and are known as *linear* polarizers. Passage of light through a linear polarizer does *not* alter the inherent integrity of the *transmitted* photons if the photons are the prevalent type of light waves that occur in nature: i.e., plane or linear waves—their axes are in a plane. A plane wave, by definition, is linearly polarized. Hence, any plane wave that successfully passes through a linear polarizer will do so with its wave function intact.

There is, however, a special category of altered photons known as *circularly* polarized photons. Instead of their electrical and magnetic field vectors traveling in linear planes (e.g., horizontal or vertical), their field vectors travel through space in unusual cork-screw or spiral configurations.

To produce a circularly polarized photon, you must pass (or reflect) a linearly polarized photon through (or off of) a special circular polarizing medium to alter the wave's original linear orientation.

Circular polarization (as opposed to linear) is unusually distinct, in part because it is not a *natural* form of light. Linearly polarized light—not circular—is the norm in nature.

Certain unusual circumstances in nature

may trigger the formation of circularly polarized light, but they are few and far between.

Linearly polarized or plane waves may make the transition to a circular orientation after striking certain surfaces at very precise angles and undergoing reflection at least twice. The reflective layers of particular beetles and crustaceans are examples.

One such example is the firefly. Of the insect's two side-by-side bodily lights, one emits circularly polarized light of the opposite rotation as that of its neighbor. Scientists believe that the light is initially generated linearly but undergoes circular rotation as it passes through the insect's external, birefringent tissue.

Certain man-made devices may also transform a linear wave to make it circular. One such device is known as a *quarter-wave plate*. This special polarizing apparatus will alter the inherent integrity (electrical and magnetic fields) of a linearly polarized photon and collapse its wave function.

From the outcome of several physics experiments, researchers have witnessed that by altering a linearly polarized photon by passing it through a circular polarizer (quarter-wave plate) will alter the photon's integrity and collapse its wave function. Of particular interest, if researchers then pass the circularly polarized photon back through a properly positioned *second quarter-wave plate*, its polarization can be returned to its original linear configuration, and its *wave function is restored*!

To repeat, if a researcher passes a linearly polarized photon through a quarter-wave plate (to make it circular) and then passes it through a second quarter-wave plate (that rotates the photon back to its original, linear orientation), its original integrity and wave function are restored—*altering even its past actions*.

After passing through the first quarter-wave plate, the light wave becomes a particle. After passing through the second quarter-wave plate, the photon transforms back to its original wave form. Not only is the photon's wave function restored with its passage through the second quarter-wave plate, but any changes in its previous behavior as a particle (as witnessed by the researchers) are also negated. For this reason, the second quarter-wave plate in such experiments is often referred to as a "reversing polarizer."

What adds to the mystery, beyond witnessing a restored photon behave retroactively, is that passing a photon's entangled twin B through a reversing polarizer (when *only* twin A has been collapsed by a quarter-wave plate) will restore the wave function of twin A! The entanglement of twin photons is an amazing and baffling phenomenon.

Thus, polarization proves itself to be quite the odd bird. The quantum world certainly does not abide by the principles predicted by Sir Isaac Newton.

You have to wonder how Einstein must be feeling. Not only does God appear to play dice with the Universe, but He plays with *loaded* dice.

Chapter 11: Dr. Higgs and Mr. Hide

The standard model of the atom predicts the existence of a field responsible for bestowing mass to all particles—creating matter. Scientists have been searching for the existence of the Higgs boson (also known as the "God particle") since 1964 when physicist Peter Higgs first suggested its speculative role in the generation of matter.

In his theory, massless "particles" (what I equate as timeless waves) interact with an energy or Higgs field. Dr. Higgs hypothesized that the kinetic energies of these massless "particles" are reduced when they encounter the field, and their lost energies are converted (via Einstein's $E = mc^2$) into mass (representing the Higgs boson).

Scientists define the Higgs field as occupying all space, and the Higgs boson (mass) represents the resultant stimulation of the Higgs field above its natural ground state, caused by its interaction with these massless "particles."

Another explanation for the Higgs field is one that I extracted from the website Wikipedia:

> The field can be pictured as a pool of molasses that "sticks" to the otherwise massless fundamental particles that travel through the field, converting them into particles with mass that form (for example) the components of atoms.[30]

What is especially apropos in the above explanation for the Higgs field is its similarity to the very role that I propose for time. I visualize time as the above-cited "molasses" (**Figure 11.1**) which transforms material waves into the mass particles that we observe, experience, and measure. Since all matter represents waves frozen in time (the Higgs field), the capture of waves within this dimension forces them into a material existence (the Higgs boson).

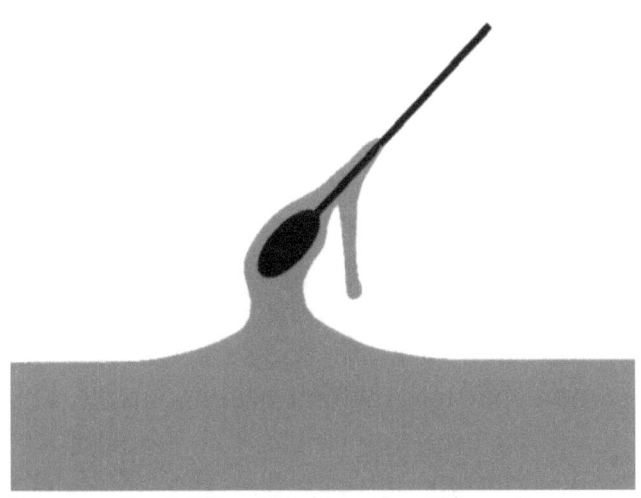

Figure 11.1 Visualizing the Higgs boson: a massless "particle" interacts with the molasses of time

Matter and time are intimately related—just as the Higgs boson relates to the Higgs field. A massless "particle" (wave) cannot interact with a Higgs boson (matter) without interacting with the Higgs field (time) as well. Consider how the non-reflective absorption of light by matter appears to be one such catalyst for transforming timeless light waves (and their entangled sources) into particles.

Science appears to be inching closer to confirming Higgs' theory. On July 4, 2012, scientists at the Large Hadron Collider (Cern, Switzerland) announced the discovery of what they believed to be the Higgs boson:

> "I think we have it," said Rolf Heuer, the director general of CERN...
>
> Dr. Heuer and others said that it was too soon to know for sure whether the new particle, which weighs in at 125 billion electron volts...fits the simplest description given by the Standard Model...
>
> Without this Higgs field, as it is known, or something like it, physicists say all the elementary forms of matter would zoom around at the speed of light, flowing through our hands like moonlight. There would be neither atoms nor life.[31]

The following day, the *Wall Street Journal* published its release:

The [Higgs] boson was proposed to fill a puzzling hole in the "standard model"…All particles ought to have zero mass—like photons, the constituents of light—and zip around space unhindered…

The newly discovered particle is…134 times as heavy as a proton.[32]

Dr. Higgs and I are in agreement. The only difference is our nomenclature. The *Higgs field is time*. When the above reporters described the elementary, massless "particles" as zooming or zipping around like light, this was merely another way of stating that they existed outside of time. I seek only to simplify the complex field of quantum mechanics by equating the Higgs field (once it is formally verified) with time. Occam's razor[33]—the scientific equivalent of KISS ("keep it simple stupid")—is apropos here.

If, on the other hand, the discovery of the Higgs is not confirmed, physicists have already offered several "Higgsless" models to explain the existence of matter. So, even among physicists, the existence of the Higgs boson is by no means mandatory.

Yet, if the Higgs boson does prove itself, then the Cern scientists and I differ only so far as semantics, and the entrapment of any wave within the unique dimension of time forces its transition to matter—just as timelessness equates with waves and the absence of mass.

Either way, I believe von Neumann was mistaken. Human consciousness is not the ingredient that collapses the wave function. It is merely the link by which we, as humans, perceive the final measurement.

Chapter 12: Conclusions

S cience suggests that an intimate relationship exists between time and matter—one cannot exist without the other. Time equates to that mysterious Higgs field—the enigmatic Medusa that transforms all waves to matter.

We have examined many concepts up to this point dealing with the existence of particle versus wave. As a recap, let us review the wave-to-particle collapse mechanisms as they applied to each of the major experiments discussed in this text. In each instance, I will identify the location of the point at which each measured wave collapsed, becoming ensnared within the time-matter continuum.

The double-slit experiment

In this experiment, a single photon passed through a partition with two slits. The slit openings represented avenues through which each light wave split into two and passed, without being absorbed by the material partition. Each half, progeny wave (entangled with its twin photon from the other slit) literally interacted (interfered) with its entangled twin, creating an interference pattern (a behavior applicable only to waves) as each successive dot accumulated on the photographic plate. The photographic plate acted as the site of wave absorption and collapse of each pair of entangled light waves. The two reunited progeny waves re-formed *a single, whole-unit photon*. Each reunited photon triggered a chemical reaction within the plate (the site of absorption and collapse) to produce a single, visible dot on the plate (the measurement).

At the moment of collapse, each photon became incorporated into the very matter of the photographic plate and its clock began to tick, signaling its transition from timeless wave to ensnared particle.

The NIST experiment

In this experiment, recall that researchers utilized a specific frequency of radio waves (a form of light) to excite the level 1 electrons of the beryllium atoms to jump to a higher electron orbit (level 2).

To measure which electrons remained behind in level 1, the researchers employed a laser (another form of light). The laser caused a selected shift in the level 1 electrons (not making the transition to level 2), which then resulted in the electrons' signature release of characteristic photons for the measurement. When these latter photons (emitted by the electrons) triggered the detector (the site of their absorption), the photons' wave functions collapsed. The collapse of the photons in turn collapsed their entangled source electrons (and probably the beryllium atoms as well), halting the progression of the experiment.

The quantum eraser experiments

Each quantum eraser experiment involved an initial light wave that split into two progeny (and entangled) photons after passing through a *transparent* crystal (wave functions emerging intact—though split).

In experiment 1, the *unmodified* set-up produced one measured response, and *modifying the end* caused another. In the modified experiment, the photons altered their paths retroactively in the middle of the experiment.

What I did not disclose earlier in this text were the fine details of the experimental set-ups.

In the *modified* experiment 1, the changes at the very end of the experiment consisted of the addition of *reversing* polarizers (or quarter-wave plates), one placed in front of each of the final measuring devices (photon detectors which represented the sites of absorption). The presence of the two reversing polarizers (placed at the *end* of the modified experiment) negated the effect of a single quarter-wave plate (circular polarizer) present in the *middle* (of both versions) of the experiment. Once again, researchers observed the integrity-restoring effect of reversing polarizers (even when placed at the end), and the twin photons altered their behaviors retroactively in the middle of the experiment (**Figure 12.1**).

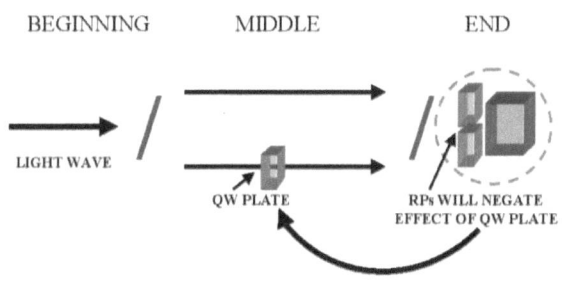

Figure 12.1 Q.E. 1: Reversing polarizers (RPs) at the end negate the effect of a quarter-wave (QW) plate in the middle.

Similarly, in revealing the details of experiment 2, the addition of a reversing polarizer in *the upper arm* of the *modified* experiment reversed the effects of a single quarter-wave plate (present in both versions of the experiment) *in the lower arm* (**Figure 12.2**).

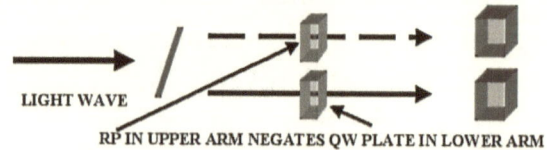

LIGHT WAVE

RP IN UPPER ARM NEGATES QW PLATE IN LOWER ARM

Figure 12.2 Q.E. 2: Upper-arm reversing polarizer (RP) negates the effect of a lower-arm quarter-wave (QW) plate.

In each of these experiments, the reversing polarizers negated the wave-destroying (wave-collapsing) effects of a single polarization filter (quarter-wave plate) in another location of the unmodified experiments. Said another way, with the addition of the reversing polarizers, the photons never lost their wave properties—despite the presence of a quarter-wave plate elsewhere in the set-up. The particle-negating effects of the reversing polarizers occurred either retroactively (experiment 1) or simultaneously but in a separate location (experiment 2).

These explanations do not make the experiments and the photons' behaviors any less baffling, however. The photons still made their changes in heretofore impossible fashions. The photons revealed that they *do* communicate (via entanglement), and they openly flaunted their unbridled disregard for both time and space.

Certainly, things can become rather confounding when discussing all the idiosyncrasies of polarization, linear and circular polarizers, reversing polarizers, double slits, refraction, and reflection—and how all these processes relate to waves and particles inside and outside of time. Waves (both EM and material) will undoubtedly continue to act with complexity enough to keep us in the dark far into the future.

The EPR experiment

In this classic experiment of wave entanglement, two photons radiated out in opposite directions from a common source (e.g., a neutral pion, or a calcium or mercury atom). The collapse and alteration of one progeny wave (e.g., photon B) by a measuring device (site of photon B's absorption) immediately collapsed its twin

(A) and conveyed mirror characteristics to both photons (**Figure 5.3**). This entanglement occurred regardless of the distance between the photons.

Transforming waves to particles

There is no reason in my mind why a light wave should behave any differently from a material (standing) wave. Size is the obvious difference between an EM wave and most material waves. Yet, if a light wave can collapse by the process of non-reflective absorption by matter, then a material wave should theoretically behave no differently.

The *time-matter continuum* should not discriminate. The material world is the medium through which we perform our daily observations and measurements. It is the medium that transforms all waves to particles.

I continue to believe that simplicity and consistency offer the truest picture of reality—Occam's razor. As such, the simplest explanation is that all waves (material and light) act alike and differ only by size. That is, the non-reflective absorption of any wave by matter will catalyze the wave (and any entangled twins and sources) to collapse.

Several scenarios follow, depicting the collapse of the wave function:

1. Light *waves* reflect from a sheet of paper. The reflected photons collapse upon a schoolgirl's retina and initiate a series of chemical reactions. The collapse of the reflected photons in turn collapses the source electrons and the atoms of the paper. The girl reads, "Mary had a little lamb."

2. A light wave triggers a photon detector to go "beep." Both the photon and its common (entangled) sources (an electron and its atom) collapse.

3. Light waves from a distant star catalyze a chemical reaction on an exposed area of film. An image appears on the film. Both the photons' and distant star's source wave functions (electrons and atoms) collapse.

4. A dark object absorbs sunlight and heats up. Both the sunlight's and Sun's source wave functions collapse.

5. In the pitch black of night, a searching hand contacts a bedside table. The table *may or may not* have already been particulate and solid before the hand touched it. Allow me to explain.

Scenario A: During the night, the table has emitted and reflected non-visible photons that have subsequently collapsed onto a host of surrounding material objects within the room. Collapse of these photons from the table have, in turn, collapsed the source electrons and atoms of the table, maintaining and reinforcing its solid state.

Scenario B: Likewise, in the *absence of any light* (visible or non-visible), any of the table's *material waves* might have chanced to interact with other materials in the bedroom. The table waves would have collapsed and taken the entire table with it. In reality however, the likelihood of the table's incredibly small waves spreading beyond the immediate confines of the table is highly unlikely—but statistically possible.

Scenario C: In the instance of the table's material waves not spreading beyond itself during the night (and in the absence of any light), then the hand contacting (and absorbing) the table's standing waves would have collapsed the table.

For most of our day-to-day reality, the photon would appear to act as the predominant intermediary that catalyzes and reinforces time's hold on our material world, but the material world cannot discriminate. All waves—whether light or material—carry the equal potential for wave-function collapse. Due to the incredibly small size of material waves, however, light more commonly performs this transformational function.

So, let us now reconsider that lone tree falling in the forest. It *will* make a sound. Von Neumann's conscious entity is not required. The continuous collapse of emitted and reflected light (visible and non-visible) from the tree (onto surrounding objects) represents the predominant mechanism that maintains the tree's materiality.

Still, it is feasible and within the realm of possibility that the tree's material waves could expand outward and become absorbed in the same manner as any entangled light waves and collapse the tree. Unlikely but possible.

Transforming particles back to waves

A much harder question is what mechanism(s) might explain the process of transforming particles and matter back to waves?

Several theories exist, none of which are totally convincing.

The predominant theory is that matter will remain as matter until enough energy is applied to the particles to catalyze their transformation back to waves.

Researchers have accomplished this repeatedly in the laboratory.

In one double-slit experiment,

researchers observed electrons behaving as waves by accelerating the particles to energies of 50,000 volts.

In several other experiments, researchers applied sufficient energy (heat) to the target materials to force their transitions from solid to gas. They then directed the energized molecules into the experimental set-up, though at relatively "slow" speeds (150m/sec in one such experiment). Although nowhere close to the speed of light, the energy appeared to be sufficient to allow the researchers to observe the molecules behaving as waves.

These early observations suggest that the addition of energy aids in the transformation of particles back to waves.

Theories do abound, however. In the NIST experiment, some scientists have considered that inertia (paradoxically) might represent one possible mechanism to explain why the beryllium electrons returned to their native wave states—but logic doesn't support this contention. True, the electrons returned to their wave configurations only when the measuring laser was off. However, during the entire experiment, radio waves—a form of radiant energy—bombarded the beryllium electrons to stimulate them to jump orbital levels.

As with the other experiments described above, I deem it more likely that the added energy (selected specifically to excite the electrons) prompted the particle-to-wave transitions in the NIST experiment.

Since speed-of-light velocities do not appear to be necessary to promote the particle-to-wave transitions discussed above, perhaps less than event-horizon gravitational forces would do likewise. Gravity is, after all, just another form of particle acceleration.

Logic (though not always appropriate in quantum physics) would suggest that any macroscopic object's large size (and small de Broglie wavelength) serves to inhibit its metamorphosis to a wave. Any large object, with photons constantly emitting and reflecting from its surface, will not likely undergo the transformation to a material wave. The entangled light waves from Einstein's tree, for instance, will be absorbed and collapsed by any nearby objects, collapsing Einstein's tree in turn.

It is amazing that recent experiments

have succeeded in measuring wave properties of molecules as large as 430 atoms and masses of 6,910 atomic mass units (AMU). The larger the particle or object, the smaller the wavelength and greater the difficulty with which to transform it to a wave (with or without experimental assistance).

To put things in perspective, consider that for a fullerene molecule (consisting of 60 carbon atoms and 720 AMU), its wavelength is 400 times smaller than the actual particle. Yet, researchers have succeeded in transforming these and much larger molecules into waves (through the selective application of energy).

For our normal, daily experiences, however, the physical properties of macroscopic objects encourage them to remain as matter. For this reason, most of our everyday conundrums involving wave-particle dualities consist mainly of atomic or subatomic particles: photons, electrons, and small atoms.

Yet, the end result is always the same: we only experience our daily reality as particles. Even when an experiment reveals that an electron or photon has acted as a wave, the measuring apparatus at the very end of the experiment has measured the result (wave interference, for example) only as particles (e.g., dots on a screen). Confusing? I won't argue.

We humans must utilize matter (e.g., a photon detector, photographic plate, cathode ray tube, or even a dot on a piece of paper) to record our observations—whether we are measuring a wave or a particle.

Conclusions

All matter, it would appear, is in a continual tug-of-war between energized wave states and absorption-induced particle states.

The Universe has not made things easy. We are trapped in the extraordinary dimension of time, and we experience only a particulate world. Meanwhile, an entirely separate realm coexists whereby matter may transform back into its natural, timeless wave state—aided through the application of energy. This energy may come in the form of acceleration, gravity, or EM radiation to name a few.

No one is sure why or how this happens. We can only speculate. Perhaps there exists an *anti-time*, which catalyzes or allows matter to return back to its original massless (wave) state—just as there exists anti-matter to matter.

Such an idea should not appear as radical as it sounds. Consider that Einstein viewed gravity as a form of acceleration. If a rocket accelerates us at a rate 32.2 ft/sec^2 (assume no gravitational fields), we would feel a force of 1 G, the same gravitational force as if we were standing on the Earth's surface.

Hence, both accelerative forces and gravity bring about similar anti-time-like effects. Although both forces cause time dilation, the effects are most recognizable at their extremes (e.g., when acceleration results in fractional light speeds or when gravity approaches the intensity of a black hole). It's as if both extreme gravity and acceleration were capable of stripping away the effects of the Higgs field (time), potentially rendering particles massless (i.e., waves).

You have to seriously consider, then, an

intrinsic link between gravity, acceleration, timelessness, waves, and the absence of mass (**Table 12.1**, right hand column, below).

Features that favor mass (particles, Higgs boson):	Features that favor masslessness (waves):
Higgs field	gravity
	acceleration
observation of finite velocities	infinite velocity (omnipresence)
time	timelessness
circular (light) polarization	linear (light) polarization
collapse of the wave function	wave function remains intact
entangled light wave absorption	light reflection, transmission
wave energy loss	particle energy gain

Table 12.1 Features favoring mass versus masslessness

Similarly, an inherent relationship must

equally exist among the opposing traits that favor mass (left hand column of **Table 12.1**).

Yet, obvious questions remain. For instance, if the Higgs field exists *everywhere* in space (as proposed by physicists), how is it that massless "particles"—i.e., waves—still exist? There should logically be no such thing as waves or massless "particles." if the Higgs field is truly everywhere. With an omnipresent Higgs field, everything should logically have mass. Hence, the Higgs field and time cannot exist everywhere—at least not in this Universe.

We know that *locations* exist where time is absent, allowing for the presence of waves. The voids between atoms appear to be such locations. Likewise, there may exist *mechanisms* which allow time not to exist—anti-time perhaps.

The very substance of matter (the Higgs boson) appears to be a co-catalyst for time (the Higgs field) to compel waves (massless "particles") to morph to their particulate states of matter.

The wave must surrender some critical and intrinsic attribute to become victim to the clutches of time. A wave's inherent integrity must become critically modified. This transition may occur, for example, when a wave is absorbed by matter and triggers an electrical, chemical, or thermal reaction—or when a photon's natural linear polarization is transformed to circular. If the natural integrity of the wave is lost, it can no longer function as a wave. It is ensnared as a constituent of the time-matter continuum—if only temporarily.

From a somewhat different perspective, these observations force us to make another conclusion: *change* occurs in our Universe only when waves—timeless energy—interact with our material reality. The ramifications may not be readily apparent, but they are profound.

If spirits exist as pure energy…if our brains emit and receive electromagnetic radiation…if God is Light…then all these energy forces are capable of creating change within our reality *only* by interacting with and becoming part of our material existence. This realization may explain the purpose and appearance of the prophets (including Christ), the angels, and the many other miracles of our sacred texts. That is, God can interact with his creations only by becoming one of us or by interacting directly within our time-matter continuum. If God and His messengers had opted to remain in their pure energy forms, we would never know of their existence.

This proves to me that God is not a passive God. Had He wanted, He could have merely observed his creations from afar and never intervened. Rather, God is an active God, performing many miracles and sending His prophets and angels to intervene regularly in our daily lives. It is no coincidence that many physicists describe physical light in divine terms. I believe that statements like "God is Light" are literal.

There is so much that we don't know. Yet, technology has come a long way to allow science to make the great strides that we have made to this point. Scientists accept the existence of a very real, external, timeless reality—far different from our own human particulate one. We know that other dimensions *DO* and must exist. We can now investigate and research at least one extraordinary, timeless dimension—the dimension of waves. Yet, through the measurement process, we alter the very intrinsic nature of this timeless reality—forcing waves into particles.

The true miracle is that humankind has progressed and evolved to the point where we have actually *accepted and proven* the existence of this alternate reality.

Still, we have far to go. We have yet to uncover a grand unified theory and to discover the nature of dark matter and energy. Yes, we have made great strides. Yet, having said that, our greatest virtue remains the recognition of our very humble and yet unique place in this vast, marvelous Universe.

Addendum: Quantum Tunneling

The absence of time as a characteristic unique to waves exists as an explanation for a previously esoteric phenomenon known as quantum tunneling.

Quantum tunneling is a phenomenon where researchers observe a particle that travels through a seemingly impossible barrier. This phenomenon is a recognized and accepted occurrence within the field of quantum physics.

Classical Newtonian physics would predict that when a particle strikes an impenetrable barrier, one of two things would happen: the particle would bounce off the barrier (reflection) or embed itself within the barrier (absorption). It could not travel through it.

Yet, the seemingly impossible phenomenon of tunneling is observed routinely. It is the accepted cause of current loss in very-large-scale integration (VLSI) electronics, producing heat and power loss: undesirable side effects so emblematic of the high–speed processors within our modern laptops, tablets, and smart phones.

To date, physicists have primarily explained tunneling using Heisenberg's uncertainty principle (which states that you cannot identify the position and momentum of a particle simultaneously). Hence, once a researcher identifies a particle's momentum, he can never know the particle's location. Thus, physicists have reasoned that the particle could theoretically exist anywhere—including impossible locations

However, quantum physics also allows for the possibility that any object (via its timeless wave function) always possesses some non–zero probability of existing literally anywhere else in the Universe. Though the probability of such a particle relocating to the other side of an impenetrable barrier may approach zero, it can never equal zero. I consider this (the particle's transition to a timeless wave) as the most likely explanation for quantum tunneling—not Heisenberg's uncertainty principle.

I judge that time is on my side and will ultimately prove to be the element that dictates either the collapse of the enigmatic wave function or its emergence from matter.

Index

A

abdominal aortic aneurysm
  Einstein's cause of death,
  5
absorption
  of light, 99
  of photons, 103
act of measuring
  in the Copenhagen
    Interpretation, 34
alternate Universes, 69
Aspect
  Alain, 17
atom
  model by Heisenberg, 8
  standard model, 85

B

Barbour
  Julian, 93
Bell's theorem, 17
Bertotti
  Bruno, 94
beryllium atoms
  in the NIST experiment,
  30
black body radiation, 8
Bohr
  Niels, 7, 56
Bohr model of the atom, 7
brain waves, 69, 71
Brownian motion, 4

C

cellphone
  waves, 24

Clauser
  John, 17
collapse of the wave
  function, 12, 18, 98
  in the NIST experiment,
  42
communication
  of photons, 125
consciousness
  causes wave-function
    collapse, 13
  collapse of the wave
    function, 12
Copenhagen Interpretation,
  6, 11, 17, 18
  early definition, 10
cosmic rays
  creating muons, 39

D

de Broglie
  Louis, 21
dimensions
  of string theory, 85
double-slit experiment, 10,
  50, 120
dreams, 69

E

$E=mc^2$, 13
Eddington
  Sir Arthur, 4
Einstein
  Albert, 2, 16
Einstein's tree, 16, 18, 34
Einstein's miracle year, 3
electric field vector, 85
  and polarization, 107

Endnotes

[1] If you travel at 186,000 mi/sec (or *any speed*), and time stops, then you theoretically can travel anywhere throughout the Universe in zero time…you can be everywhere at once (or omnipresent)!

[2] Itano, Wayne M., Heinzen, D.J., Bollinger, J.J., and Wineland, D.J. "Quantum Zeno effect." *Physical Review A.* 41:5(1 Mar. 1990): 2295-2300.

[3] Baumann, T. Lee. *God at the Speed of Light: The Melding of Science and Spirituality.* Virginia Beach, VA: A.R.E. Press, 2001.

[4] "Werner Heisenberg." Wikipedia.org, http://en.wikipedia.org/wiki/Heisenberg (25 April 2012) 11 April 2012.

[5] "Niels Bohr." Wikipedia.org, http://en.wikipedia.org/wiki/Niels_Bohr (25 April 2012) 24 April 2012.

[6] Planck's constant is the value 6.63×10^{-34} joule $\cdot$ sec.

[7] Herbert, Nick. *Quantum Reality.* New York: Anchor Books, 1985, p. 148.

[8] Sternglass, Ernest J. *Before the Big Bang.* New York: Four Walls Eight Windows, 1997, p. 53.

[9] Baumann, T. Lee. *Matter to Mind to Consciousness: Anatomy of the E.L.F.* North Charleston, SC, 2011.

[10] Itano, Wayne M., Heinzen, D.J., Bollinger, J.J., and Wineland, D.J. "Quantum Zeno effect." *Physical Review A.* 41:5(1 Mar. 1990): 2295-2300.

[11] "Atomic orbital." Wikipedia.org, http://en.wikipedia.org/wiki/Atomic_orbital (12 April 2012) 11 April 2012.

[12] Gribbin, John. *Schrodinger's Kittens and the Search for Reality.* New York: Back Bay Books, 1995 p. 132.

[13] Although one recent publication cited evidence for a faster-than-light particle, the results were subsequently retracted as erroneous.

[14] In contrast, weightlessness and lack of motion speed the passage of time.

[15] Gribbin, John. *Schrodinger's Kittens and the Search for Reality,* op. cit., pp. 134- 35.

[16] To make things even more bizarre, Einstein's special theory made another striking conclusion—also proven. An

observer traveling at a fractional light speed will observe his spatial dimensions to shrink and his mass to increase—in addition to his clock slowing.

[17] Gribbin, John. *Schrodinger's Kittens and the Search for Reality*. New York: Back Bay Books, 1995, p. 79

[18] Gribbin, John. *In Search of Schrodinger's Cat*. New York: Bantam Books, 1984, p. 191

[19] Goswami, Ph.D., Amit, with Richard E. Reed and Maggie Goswami. *The Self-Aware Universe*. New York: Jeremy P. Tarcher / Putnam, 1993, p. 73.

[20] Feynman, R.P., Leighton, R.B., and Matthew Sands. *The Feynman Lectures on Physics*. Reading, Massachusetts: Addison-Wesley Publishing Company, 1963, p. 37-2.

[21] Perhaps this is why we must leave our bodies behind at the time of death.

[22] Physicists have derived the concept of light's omnipotence from several factors: the mathematical technique of *renormalization*, the history surrounding the *ultraviolet catastrophe*, and the

accepted observation that photons can travel infinite distances.

[23] *Merriam Webster's Collegiate Dictionary*. Springfield, MA: Merriam-Webster, Inc., 1993.

[24] "Time." Wikipedia.org, http://en.wikipedia.org/wiki/Time (6 July 2012) 6 July 2012.

[25] *Merriam Webster's Collegiate Dictionary*. Springfield, MA: Merriam-Webster, Inc., 1993.

[26] The concept of proton decay states that all matter in the Universe will eventually decompose to nothing more than just protons and electrons. The proton is projected to have a half-life of about 10^{36} years, and each will subsequently decay into a positron and a neutral pion. The positron annihilates with its antiparticle, the electron, creating 2 photons. Further, the neutral pion spontaneously combusts into 2 gamma ray photons. The end result is the existence of only Light within the Universe.

[27] Barbour, J.B. "The Emergence of Time and Its Arrow from Timelessness." Halliwell, J. J., Perez-Mercader, J., Zurek,

W.H., editors. *The Physical Origins of Time Asymmetry*. Cambridge, UK, Cambridge University Press, 1996, p. 409
[28]"J. M. E. McTaggart." Wikipedia.org, http://en.wikipedia.org/wiki/J._M._E._Mc Taggart (6 May 2012) 2 May 2012.
[29] Note of clarification: I should point out that the collapse of the wave function has nothing to do with either nuclear fusion (e.g., the mechanism behind the hydrogen bomb) or, conversely, fission (the atomic bomb). Although I keep referring to Einstein's $E = mc^2$, I do so only to remind you that energy and matter are interchangeable. I am not inferring that any *significant* energy exchange or release is part of the wave-function collapse—other than, perhaps, at the photonic level. The energy release from splitting the atom, for instance, is not what I am discussing. The act of measuring or observing the molecules of Einstein's tree does not involve either the fusion or fission of atoms. The tree's atoms have, rather, undergone a metamorphosis of sorts, from waves to particles.

---

[30] "Higgs Boson." Wikipedia.org, https://en.wikipedia.org/wiki/Higgs_boson (3 June 2012) 30 May 2012

[31] Overbye, Dennis. "A New Particle Could Be Physics' Holy Grail." NYTimes.com, https://www.nytimes.com/2012/07/05/science/cern-physicists-may-have-discovered-higgs-boson-particle.html?_r=1&pagewanted=all (4 July 2012) 4 July 2012.

[32] Naik, Gautam. "Discovery May Help Tell Universe's Secrets." *The Wall Street Journal*, 5 July 2012: p. A3.

[33] Occam's razor is "the scientific and philosophic rule that…the simplest of competing theories be preferred to the more complex" [*Merriam Webster's Collegiate Dictionary*. Springfield, MA: Merriam-Webster, Inc., 1993]. In other words, if you can explain a concept completely in a few words (think of Einstein's $E = mc^2$), then that explanation is likely to be more accurate (and certainly preferred) to one involving several hundred words.